T0184000

Lecture Notes in Bioinformatics 10834

Subseries of Lecture Notes in Computer Science

More information about this series at http://www.springer.com/series/5381

Massimo Bartoletti · Annalisa Barla
Andrea Bracciali · Gunnar W. Klau
Leif Peterson · Alberto Policriti
Roberto Tagliaferri (Eds.)

Computational Intelligence Methods for Bioinformatics and Biostatistics

14th International Meeting, CIBB 2017
Cagliari, Italy, September 7–9, 2017
Revised Selected Papers

 Springer

Editors
Massimo Bartoletti ⓘ
University of Cagliari
Cagliari, Italy

Annalisa Barla ⓘ
University of Genova
Genoa, Italy

Andrea Bracciali ⓘ
University of Stirling
Stirling, UK

Gunnar W. Klau ⓘ
Heinrich-Heine-University Düsseldorf
Düsseldorf, Germany

Leif Peterson ⓘ
Houston Methodist Research Institute
Houston, TX, USA

Alberto Policriti ⓘ
University of Udine
Udine, Italy

Roberto Tagliaferri ⓘ
University of Salerno
Fisciano, Italy

ISSN 0302-9743 ISSN 1611-3349 (electronic)
Lecture Notes in Bioinformatics
ISBN 978-3-030-14159-2 ISBN 978-3-030-14160-8 (eBook)
https://doi.org/10.1007/978-3-030-14160-8

Library of Congress Control Number: 2019932170

LNCS Sublibrary: SL8 – Bioinformatics

This Springer imprint is published by the registered company Springer Nature Switzerland AG
The registered company address is: Gewerbestrasse 11, 6330 Cham, Switzerland

Preface

The 14th annual edition of the International Meeting on Computational Intelligence methods for Bioinformatics and Biostatistics (CIBB 2017) built upon the tradition of the CIBB conference series and provided a multi-disciplinary forum open to researchers interested in the application of computational intelligence, in a broad sense, to open problems in bioinformatics, biostatistics, systems and synthetic biology, medical informatics, as well as computational approaches to life sciences in general.

In line with the spirit of CIBB, the 2017 meeting brought together researchers from different communities who address problems from different, but connected and often overlapping, perspectives. CIBB 2017 tackled the difficult task of bridging different backgrounds by providing an inclusive venue to discuss advances and future perspectives in different areas. It also fostered interaction between theory and practice, addressing both the theories underpinning the methodologies used to model and analyze biological systems, the practical applications of such theories, and the supporting technologies. Accordingly, participants at CIBB 2017 came from mathematical, computational, and medical backgrounds and institutions, both from academia and the private sector, offering collaboration opportunities and novel results in the areas of computational life sciences.

CIBB 2017 also offered a view on emerging and strongly developing trends and future opportunities at the edge of mathematics as well as computer and life sciences, such as synthetic biology, statistical investigation of genomic data, and applications to the understanding of complex diseases, such as cancer, and therapy opportunities. Along these lines, three keynote speakers, prominent scholars in their fields, presented the latest advances of their research within the context of their area of interest, and provided insights into open problems and future directions of general interest for the field. While papers in the main conference track addressed a rich set of open problems at the forefront of current research, the conference hosted three further special sessions on specific themes: synthesis of artificial cells by combining bio-interfaces engineering and systems biology modeling, modeling and simulation methods for systems biology and systems medicine, and molecular communication. Researchers from Europe, Asia, and America attended the conference.

CIBB 2017 was made possible by the efforts of the Organizing, Program, and Steering Committees and by the support of sponsors and participants. CIBB 2017 was held in Cagliari, Italy, during September 7–9, 2017. Overall, 44 contributions were submitted for consideration to CIBB 2017, amongst which 33 were invited for an oral presentation at the conference, after a first round of reviews (at this stage, each paper received an average of 3.25 reviews from the Program Committee and about 13 additional referees). Following the conference, selected papers were invited for further submission, after feedback and discussion from the conference. This volume collects the papers that were accepted after a further round of reviews (2.5 for each paper, on average).

From 2004 to 2007, CIBB had the format of a special session of larger conferences, namely, WIRN 2004 in Perugia, WILF 2005 in Crema, FLINS 2006 in Genoa, and WILF 2007 in Camogli. Given the great success of the special session at WILF 2007 that included 26 strongly rated papers, the Steering Committee decided to turn CIBB into an autonomous conference starting with the 2008 edition in Vietri. The following editions in Italian venues were held in Genoa (2009), Palermo (2010), and Gargnano (2011). Until 2012, CIBB meetings were held annually in Italy with an increasing number of participants. CIBB 2012 was the first edition organized outside Italy, in Houston, then in Nice, France (2013), Cambridge, UK (2014), Naples, Italy (2015), and Stirling, UK (2016).

A rigorous peer-review selection process is applied every time to ultimately select the papers included in the program of the conference, in the conference proceedings published in the LNBI-LNCS book series by Springer, and in some cases, selected papers were published in special issues of well-qualified international journals, such as *BMC Bioinformatics*.

December 2018

Annalisa Barla
Massimo Bartoletti
Andrea Bracciali
Gunnar W. Klau
Leif Peterson
Alberto Policriti
Roberto Tagliaferri

Organization

Program Committee

Antonino Abbruzzo	University of Palermo, Italy
Marco Antoniotti	University of Milano-Bicocca, Italy
Sansanee Auephanwiriyakul	Chiang Mai University, Thailand
Massimo Bartoletti	University of Cagliari, Italy
Gilles Bernot	University of Nice Sophia Antipolis, France
Daniela Besozzi	University of Milano-Bicocca, Italy
Andrea Bracciali	University of Stirling, UK
Stefan Canzar	Gene Center, LMU, Germany
Giulio Caravagna	Institute of Cancer Research, UK
Paolo Cazzaniga	University of Bergamo, Italy
Davide Chicco	Peter Munk Cardiac Centre, Canada
Angelo Ciaramella	University of Naples Parthenope, Italy
Luisa Cutillo	SITraN, University of Sheffield, UK
Chiara Damiani	University of Milano-Bicocca, Italy
Angelo Facchiano	CNR, Istituto di Scienze dell'Alimentazione, Italy
Alessandro Filisetti	Explora Biotech Srl, Italy
Enrico Formenti	University of Nice Sophia Antipolis, France
Christoph M. Friedrich	University of Applied Science and Arts Dortmund, Germany
Yair Goldberg	University of Haifa, Israel
Alex Graudenzi	University of Milano-Bicocca, Italy
Sean Holden	University of Cambridge, UK
Gunnar W. Klau	Heinrich-Heine-University Düsseldorf, Germany
Johannes Köster	University of Duisburg-Essen, Germany
Paola Lecca	University of Trento, Italy
Hassan Mahmoud	University of Genoa, Italy
Anna Marabotti	University of Salerno, Italy
Roberto Marangoni	University of Pisa, Italy
Tobias Marschall	Saarland University/Max Planck Institute for Informatics, Germany
Giancarlo Mauri	University of Milano-Bicocca, Italy
Fabio Mavelli	University of Bari "Aldo Moro", Italy
Pasquale Palumbo	IASI-CNR, Italy
Dario Pescini	University of Milano-Bicocca, Italy
Leif Peterson	Houston Methodist Research Institute, USA
Alberto Policriti	University of Udine, Italy
Vilda Purutçuoğlu	Middle East Technical University, Turkey

Davide Risso	Cornell University, USA
Riccardo Rizzo	ICAR-CNR, Italy
Andrea Roli	University of Bologna, Italy
Federico Rossi	University of Salerno, Italy
Stefano Rovetta	University of Genova, Italy
Antonino Staiano	University of Naples Parthenope, Italy
Pasquale Stano	University of Salento, Italy
Francesco Stingo	MD Anderson Cancer Center, USA
Roberto Tagliaferri	University of Salerno, Italy
Paolo Tieri	Consiglio Nazionale delle Ricerche, Italy
Filippo Utro	IBM T. J. Watson Research, USA
Alfredo Vellido	Universitat Politècnica de Catalunya, Spain
Paweł P. Łabaj	Boku University Vienna, Austria

Additional Reviewers

Alqassem, Israa
Angaroni, Fabrizio
Femminella, Mauro
Graudenzi, Alex
Kohvaei, Parastou

Nouri, Nima
Pucciarelli, Sandra
Ramazzotti, Daniele
Rautiainen, Mikko
Rossi, Nicolò

Contents

An Open-Source Tool for Managing Time-Evolving Variant Annotation

Ilio Catallo[1]([✉]), Eleonora Ciceri[1], Stefania Stenirri[1], Stefania Merella[2], Alberto Sanna[1], Maurizio Ferrari[2,3,4], Paola Carrera[2,3], and Sauro Vicini[1]

[1] e-Services for Life and Health, San Raffaele Scientific Institute, Milan, Italy
{catallo.ilio,ciceri.eleonora,stenirri.stefania,sanna.alberto, vicini.sauro}@hsr.it
[2] Clinical Molecular Biology Laboratory, San Raffaele Scientific Institute, Milan, Italy
{merella.stefania,ferrari.maurizio,carrera.paola}@hsr.it
[3] Unit of Genomics for Human Disease Diagnosis, San Raffaele Scientific Institute, Milan, Italy
[4] Chair of Clinical Pathology, Vita-Salute San Raffaele University, Milan, Italy

Abstract. During the past decade, genomics has been drawing more and more attention, thanks to the introduction of fast and accurate sequencing strategies. Accumulation of data is fast and the amount of information to be managed and integrated is snowballing. While new variants are discovered every day, we still do not know enough about the human genome to have a final understanding of all the implications that they could have from a clinical point of view. When inherited diseases are considered, variants clinical classification may change over time, in relation to new discoveries. In this scenario, software solutions that help operators in the analysis and maintenance of constantly changing genomic data are relevant in the field of modern molecular medicine. In this paper we present GLIMS (short for *Genomics Laboratory Information Management System*), an open-source laboratory information management system for genomic data that allows to deal with time-evolving variant annotations. This solution answers to the need of genomic laboratories to keep up with their knowledge about variants and annotations, so as to provide patients with up-to-date reports. We illustrate the architecture of GLIMS modules that are in charge of keeping the database of variants updated and reclassifying patients' variants. Then, we demonstrate (via the use of GLIMS) that variant clinical classifications are changing rapidly even in ClinVar, one of the most known and cited genomic databases, thus underlining the need for a tool that tracks changes over time.

1 Scientific Background

A *genome* is the genetic material of an individual. The genetic instructions it contains are used in the growth, development, functioning and reproduction of individuals, and define one's phenotype Genomes contain *genes*, that is, regions

© Springer Nature Switzerland AG 2019
M. Bartoletti et al. (Eds.): CIBB 2017, LNBI 10834, pp. 1–8, 2019.
https://doi.org/10.1007/978-3-030-14160-8_1

of DNA that encode specific functions. Genes can acquire mutations in their sequence of nucleotides, leading to different *variants* in the population. Every variant comes with a set of *genomic annotations*, which state its semantics (e.g., specifying whether it is associated with an increased probability of developing a pathology) and its biological structure. Over the years, scientists have published several archives of genomic variants and annotations. A well-known example is ClinVar [1], a freely available archive of clinically significant relationships among human variations and phenotypes. Such archives are not limited to specific pathologies and are constantly updated to reflect the knowledge that researchers acquire over time, providing benefits on two main aspects. First, the constant update permits the gathering of new information about Variants of Unknown Significance (VUS), which are known to change very frequently towards a clear pathological significance [6]. As a matter of fact, an uncertain finding can be frustrating for clinicians and patients alike, who may decide for drastic treatment measures (e.g., surgical decisions) only in the name of the persistent fear of contracting a disease [7]. Second, since multiple sources of genomic information may differently interpret the clinical significance of a variant [8], archives of genomic variants often include errors and misinterpretations [9] and present classification inconsistencies (even for well-studied genomic panels [10]). In this regard, only a continuous update guarantees that exams will be carried out against the most coherent annotations.

Thus, it is vital for laboratories to: *(i)* keep their knowledge about variant classifications updated with respect to the current literature; *(ii)* track changes in variant classifications in order to identify which patients' past genetic results are in need of a review. Several published works already outlined the necessity of notifying patients when their genetic test clinical significance changes as a consequence of a variant reclassification, arguing about the clinical impact and ethical duties of such actions [11,12]. From a technical point of view, the introduction of an automatic solution seems to be the best option at scale, as the exponential growth of interest in the genomic field has brought to the production of an unprecedented mass of genomic data and tests, and laboratories would incur in high costs if they had to manually reclassify patients' exams. Some works in the state of the art already presented automatic solutions that allow laboratories to update database variants and genetic test reports upon reclassification of variants [13–15]. Still, these solutions are mainly proprietary, and work with non-standard (if not unknown) data format. For instance, in [16] the authors presents a pipeline based on IBM Watson (which is not open-source). Moreover, their scope is limited to targeted exons and tumor-related data.

In this paper, we present GLIMS (short for *Genomics Laboratory Information Management System*), an open-source laboratory information management system for genomic data[1], that, in addition to automatizing alignment and annotation tasks, also provides reclassification capabilities for time-evolving variant

---

[1] As of this writing, we are perfecting an imminent release on a publicly-available repository. In the meantime, please reach the corresponding author for the source code.

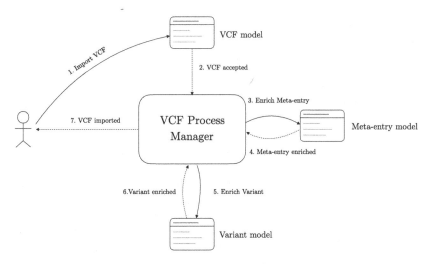

Fig. 1. The sequence of operations needed for enriching the internal database. The upload of a VCF file triggers the *VCF Process Manager* component, which takes care of orchestrating the importing procedure. Lines in the input VCF are converted to *Enrich Meta-entry* and *Enrich Variant* commands, which are then directed to the proper *Meta-entry* and *Variant* models

annotations. GLIMS promotes interoperability between systems by adopting the well-known Variant Call Format (VCF) standard as the exporting and importing format for genomic variants.

2   Materials and Methods

GLIMS takes care of two important steps in the analysis of Next-Generation Sequencing (NGS) data, that is, sequence alignment and variant annotation[2], in a privacy-compliant cloud environment (in order to reduce the burden of such expensive computations). On top of that, GLIMS also provides two more distinctive functionalities, namely, that of *(i)* growing an internal database of variants encountered in patients, which can be exported in VCF format; and *(ii)* supporting periodic reclassification of variants so as to update patients' record whenever there is a change in their variant annotations.

The management of the internal database of variants and the annotation process in GLIMS are supported by the usage of the well-known VCF file format. A VCF file is a text file used for storing gene sequence variations, and it has been widely used in the last years, with the support of large-scale DNA sequencing projects such as ExAC [2] and 1000 Genomes [3]. The diffusion of the VCF format among the most known genomic databases allows GLIMS to import

---

[2] Alignment and annotations may be carried out with an array of off-the-shelf tools. While at present GLIMS uses `bwa` [4] and `SnpSift` [5], it remains agnostic about the particular technological choice.

Table 1. Sequence of events stored into GLIMS in a given moment in time

Event	Timestamp
VariantEnriched(id=4c0e94, info=CLNDBN, ...)	2017-04-18T17:14:39.097Z
VariantEnriched(id=4c0e94, info=CLNSIG, ...)	2017-01-10T11:10:28.023Z
VariantCreated(id=4c0e94, chrom=chr17, ...)	2016-03-27T16:37:02.862Z

any information they provide, and keep it updated whenever a new version of such databases is released. On the other way around, GLIMS makes it easy for biologists to share custom annotations they may have recorded into the system, since the platform already stores them in a standard format. A VCF file encodes variant characteristics by subdividing the information in two sections: meta-entries and variants. A *meta-entry* provides metadata describing the file content (e.g., the semantics of a variant annotation). A *variant* provides the description of a genomic variant, defining its characteristics (e.g., chromosome, position on the chromosome, expected nucleotide according to the reference genome, found alternative) and decorating it with its annotations.

The process of enriching the internal database with new variants (and their related annotation) is depicted in Fig. 1. As shown, the act of importing a VCF is organized as a sequence of *commands* and *events*. When a user asks for a VCF file to be imported, GLIMS stores her request as a new *VCF* model, which in turn causes a *VCF accepted* event to be fired[3]. Such an event is captured by the *VCF Process Manager* component, which initiates and thereupon orchestrates the enriching process. At this point, depending on their position and content, lines in the input VCF are converted to *Enrich Meta-entry* and *Enrich Variant* commands, which are then directed to the proper *Meta-entry* and *Variant* models.

GLIMS stores every event happening in the system. An example of possible events is reported in Table 1. As shown, each event describes how the system has been altered as a consequence of its occurrence. This means that the state of any model can be reconstructed by simply re-applying what happened in the past. In this respect, the presence of a *timestamp* allows every model to be restored to a specific moment in time.

A pictorial representation of the annotation process is presented in Fig. 2. This process can be used for creating a new exam, as well as for its periodic reclassification, in that reclassification can be seen as a sequence of two annotations on the same exam. Like the enriching process, the interaction between components is driven by commands and events. A user triggers the annotation of a patient's exam by issuing an *Annotate* command. The *Annotator* component reacts to such a request by first making a local copy of the internal database, fetching an up-to-date version of each requested remote database, and then

---

[3] It is important to note that *VCF accepted* events are always fired asynchronously, as the process of importing even small-sized VCF files has to be considered a long-running operation.

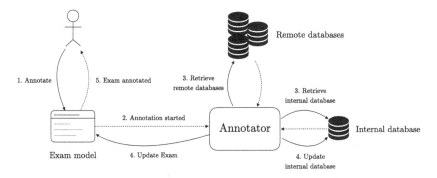

Fig. 2. The sequence of operations needed for the annotation of a patient's exam. Upon invocation, the *Annotator* component retrieves both the internal and the remote databases and then proceeds with the annotation. Since the updated exam may contain novel information, the *Annotator* also takes care of updating the internal database

Table 2. Dataset: ClinVar releases

2014	$2016^{(1)}$	$2016^{(2)}$	$2017^{(1)}$	$2017^{(2)}$
February 2014	January 2016	November 2016	February 2017	April 2017

proceeding with the actual annotation. Upon completion, the output VCF may contain novel information with respect to both the internal database and the current state of the exam. Therefore, the *Annotator* takes care of submitting the VCF to the internal database, as well as updating the *Exam* model of interest. Finally, the user is notified with an *Exam annotated* event, which informs her (e.g., via a notification through the user interface) about relevant changes in the annotation of patients' variants.

As anticipated, the crucial aspect for the *Annotator* component is to annotate the patient's variants against the appropriate version of the internal database. To this end, the exam model maintains a list of references to every database that has ever been used for its annotation. When performing a reclassification, the *Annotator* component can therefore instruct the internal database to provide only those variants that have been subject to change since the last reclassification.

3 Results

In this section we demonstrate the need of an automatic solution, such as the one proposed by GLIMS, by evaluating the variability of genomic variants and annotations contained in ClinVar.

We retrieved five ClinVar GRCh37 (human `hg19`) releases (Table 2), retained only the variants belonging to the `BRCA1`/`BRCA2` genes and finally imported the resulting VCF files in GLIMS. For each considered release, we then computed: *(i)* the number of new variants introduced with respect to release 2014; *(ii)* the number of variants whose clinical significance (`CLNSIG`) annotation or variant disease name (`CLNDBN`) annotation changed with respect to previous releases.

Figure 3 reports the number of new variants added to ClinVar since release 2014. It is shown that the volume of data in ClinVar continues to grow, gaining up to 360% of the initial volume in two years. This underlines the necessity of maintaining the knowledge on variants constantly updated, so that when a patient requires a new genomic test, her variants can be correctly interpreted at the best of researchers' knowledge.

Figure 4 reports the number of release-to-release changes in `CLNSIG` or `CLNDBN` annotations. It is shown that changes happen frequently also for variants that are already known to the scientific community, updating up to 75% of variant classifications with respect to what was reported in the previous release. This highlights the need of updating periodically the outcomes for already performed genomic tests, so that if an important change on the clinical significance of one's variants is found, then she can be timely notified with an updated report.

These results are further confirmed if the updated variants are analyzed in detail, as in the case of the four variants presented in Table 3. In all these cases, the provided update would impact on the final interpretation on one's variants, and thus this links to the necessity of providing patients with an updated report upon reclassification. Reclassification of variants may have an impact on patients' follow-ups (in terms of surveillance, preventive or pharmacological treatments) as well as on the evaluation of risk for their relatives.

The above discussion highlights how discovering changes in annotations (and specifically for changes in annotations conveying clinical significance) is a pressing concern for diagnostic laboratories. Still, its manual execution remains a time-consuming, infeasible tasks when the genomic data at hand are too large. In this respect, GLIMS's automatic reclassification of exams may be decisive to the capability of laboratories to provide timely and relevant information to patients.

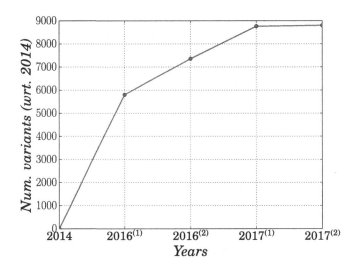

Fig. 3. New variants added to ClinVar since release 2014

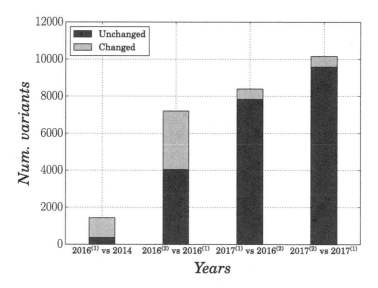

Fig. 4. Variants with changed and unchanged `CLNSIG` or `CLNDBN` annotations

Table 3. Variants evolving over time (human reference genome release: GRCh37/`hg19`). The reported HGVS nomenclature is the standard for the description of sequence variations.

Variants (HGVS nomenclature)	Gene	Position	CLNSIG	
			2014	$2017^{(2)}$
NM_007294.3(BRCA1) : c.190T > G	BRCA1	41258495	VUS	Pathological
NM_007294.3(BRCA1) : c.3119G > A	BRCA1	41244429	VUS	Benign
NM_000059.3(BRCA2) : c.1889C > T	BRCA2	32907504	VUS	Benign
NM_000059.3(BRCA2) : c.5428G > A	BRCA2	32913920	Pathological	VUS

4   Conclusion

In this paper we presented GLIMS, an open-source laboratory information management system for genomic data on germline mutations, and focused our attention on two distinctive functionalities: the capability of maintaining an evolving database of variants over time, and its support to the periodic reclassification of patients' variants. These two functionalities, when combined, offer the possibility of providing patients with fresh and up-to-date information about their genomic variants, which are known to be always changing and in need of refinements (as also proven by our pilot study on the `BRCA` genes in the ClinVar database). The pilot will be expanded including an extensive analysis of the tool with larger panels, in cooperation with the Clinical Molecular Biology Laboratory at San Raffaele Hospital. This project may include other genomic variations such as Copy Number Variations (CNV), somatic mutations and data from other

OMICS studies (e.g., epigenetics, proteomics, transcriptomics). Machine learning techniques may be developed for the automatic classification of VUS clinical significance and the computation of genotype-phenotype correlation. These applications would be extremely important and helpful not only for the integration of data, correlation to the clinical phenotype and formulation of hypotheses, but also in the process of harmonization and standardization of protocols. Heterogeneity of the conclusions drawn by different operators is in fact a variable with an important impact on the classification of genetic variants as well as on the clinical management.

Acknowledgments. This work is partially funded by the EU H2020 Framework Programme under project WITDOM (project no. 644371).

References

1. ClinVar. https://www.ncbi.nlm.nih.gov/clinvar/
2. ExAC. http://exac.broadinstitute.org/
3. Genomes. http://www.internationalgenome.org/
4. Burrows-Wheeler Aligner. http://bio-bwa.sourceforge.net/
5. SnpEff and SnpSift. http://snpeff.sourceforge.net/
6. Narravula, A., Garber, K.B., Askree, S.H., Hegde, M., Hall, P.L.: Variants of uncertain significance in newborn screening disorders: implications for large-scale genomic sequencing. Genet. Med. **19**, 77 (2017)
7. Murray, M.L., Cerrato, F., Bennett, R.L., Jarvik, G.P.: Follow-up of carriers of BRCA1 and BRCA2 variants of unknown significance: variant reclassification and surgical decisions. Genet. Med. **13**, 998 (2011)
8. Garber, K.B., Vincent, L.M., Alexander, J.J., Bean, L.J., Bale, S., Hegde, M.: Reassessment of genomic sequence variation to harmonize interpretation for personalized medicine. Am. J. Hum. Genet. **99**, 1140–1149 (2016)
9. Salgado, D., Bellgard, M.I., Desvignes, J.P., Béroud, C.: How to identify pathogenic mutations among all those variations: variant annotation and filtration in the genome sequencing era. Hum. Mutat. **37**, 1272–1282 (2016)
10. Lincoln, S.E., et al.: Consistency of BRCA1 and BRCA2 variant classifications among clinical diagnostic laboratories. JCO Precis. Oncol. **1**, 1–10 (2017)
11. Otten, E., et al.: Is there a duty to recontact in light of new genetic technologies? A systematic review of the literature. Genet. Med. **17**, 668 (2015)
12. Dheensa, S., et al.: A 'joint venture' model of recontacting in clinical genomics: challenges for responsible implementation. Eur. J. Med. Genet. **60**, 403–409 (2017)
13. Bean, L.J., Tinker, S.W., da Silva, C., Hegde, M.R.: Free the data: one laboratory's approach to knowledge-based genomic variant classification and preparation for emr integration of genomic data. Hum. Mutat. **34**, 1183–1188 (2013)
14. Aronson, S.J., Clark, E.H., Varugheese, M., Baxter, S., Babb, L.J., Rehm, H.L.: Communicating new knowledge on previously reported genetic variants. Genet. Med. **14**, 713 (2012)
15. Wilcox, A.R., et al.: A novel clinician interface to improve clinician access to up-to-date genetic results. J. Am. Med. Inform. Assoc. **21**, e117–e121 (2014)
16. Patel, N.M., et al.: Enhancing next-generation sequencing-guided cancer care through cognitive computing. Oncologist **23**, 179–185 (2018)

Extracting Few Representative Reconciliations with Host Switches

Mattia Gastaldello[1,2], Tiziana Calamoneri[1(✉)], and Marie-France Sagot[2]

[1] Computer Science Department, Sapienza University of Rome, Rome, Italy
calamo@di.uniroma1.it

[2] Inria Grenoble - Rhône-Alpes & Université de Lyon, Université Lyon 1; CNRS,
UMR5558, 69000 Lyon, France

In memory of Mattia

Abstract. Phylogenetic tree reconciliation is the approach commonly used to investigate the coevolution of sets of organisms such as hosts and symbionts. Given a phylogenetic tree for each such set, respectively denoted by H and S, together with a mapping ϕ of the leaves of S to the leaves of H, a reconciliation is a mapping ϱ of the internal vertices of S to the vertices of H which extends ϕ with some constraints.

Given a cost for each reconciliation, a huge number of most parsimonious ones are possible, even exponential in the dimension of the trees. Without further information, any biological interpretation of the underlying coevolution would require that all optimal solutions are enumerated and examined. The latter is however impossible without providing some sort of high level view of the situation. One approach would be to extract a small number of representatives, based on some notion of similarity or of equivalence between the reconciliations.

In this paper, we define two equivalence relations that allow one to identify many reconciliations with a single one, thereby reducing their number. Extensive experiments indicate that the number of output solutions greatly decreases in general. By how much clearly depends on the constraints that are given as input.

Keywords: Cophylogeny · Reconciliations · Equivalence relation

1 Scientific Background

Given a directed binary tree T, we denote by $V(T)$ and $A(T)$ the set of its vertices and arcs, respectively. Given $v \in V(T)$, we denote by $p(v)$ its *parent* and by $s(v)$ its (unique) *sibling*.

This work has been partially supported by *Italian-French University* (Project "Algorithms and Models for Solving Complex Problems in Biology") and *Sapienza University of Rome* (Project "Combinatorial structures and Algorithms for Co-Phylogeny Problems"). The tables are filled also with the aids of the computing facilities of the CC LBBE/PRABI.

M. Bartoletti et al. (Eds.): CIBB 2017, LNBI 10834, pp. 9–18, 2019.
https://doi.org/10.1007/978-3-030-14160-8_2

Given two vertices $u, v \in V(T)$, u is an *ancestor* of v, denoted by $u \succeq_T v$, if either $u = v$ or there exists a directed path from u to v. If either $u \succeq_T v$ or $v \succeq_T u$, then we say that they are *comparable*. We say that u and v are *incomparable* if there is not a directed path between u and v.

If $u \succeq_T v$, we denote by $path_T(u, v) = (t_1, \ldots, t_j)$ the (unique) ordered sequence of vertices of T traversed along the directed path from u to v. Of course, $t_1 = u$ and $t_j = v$.

A *phylogenetic tree* T is a leaf-labelled rooted binary tree that models the evolution of a set of taxa (placed at the leaves) from their most recent common ancestor (placed at the root). The internal vertices of the tree correspond to the speciation events.

The model of host-symbiont evolution we rely on in this paper is the event-based one [1,15]. Let H and S be the phylogenetic trees for the host and symbiont species, respectively. A function ϕ is defined from the leaves of S to the leaves of H that indicates the association between currently living host and symbiont species.

A *reconciliation* ϱ is a function from the set of internal vertices of S to the set of vertices of H that extends the mapping ϕ of the leaves under some constraints. Note that each internal vertex of S can be associated to an event among: *cospeciation* (when both the symbiont and the host speciate), *duplication* (when the symbiont speciates but not the host) and *host switch* (when the symbiont speciate and one of its children is associated to an incomparable host), while each arc (u, v) of S is associated to a certain number of *loss* events $l_{(u,v)} \geq 0$ that is equal to the length of $path_H(\varrho(u), \varrho(v))$ if $\varrho(u) \succeq_H \varrho(v)$. It is therefore possible to associate to each reconciliation ϱ a vector $E_\varrho = \langle e_c, e_d, e_s, e_l \rangle$ [2], that we call *event vector*, where e_c, e_d, e_s and e_l denote the number of cospeciations, duplications, host switches and losses, respectively, that are in ϱ.

Given a vector $C = \langle c_c, c_d, c_s, c_l \rangle$ of real values that correspond to the cost of each type of event, the most parsimonious (or optimal) reconciliations are the ones that minimise the total cost, *i.e.* that minimise $cost(\varrho) = \sum_{i \in \{c,d,s,l\}} e_i c_i$. Note that it is usual to assume $c_c < c_d$ and $c_l > 0$; in the following we adopt these assumptions. We denote by $\mathcal{R}(H, S, \phi, C)$ the set of all optimal reconciliations from the tree S to the tree H whose leaves are connected by means of the mapping ϕ, and in which the costs of the events are given by C.

In the context of gene-species associations, our model is known as DTL and has been extensively studied (see, for example, [1,6,9,14,15]). The *least common ancestor* of a set W of vertices of T is the lowest vertex of T that is an ancestor of every vertex in W. The so-called *LCA mapping* of a symbiont vertex $p \in S$ [7] is defined as the least common ancestor in H of all host species where a symbiont descended from p has been mapped, inductively computed with ϕ as base of the induction. It is known that the LCA mapping induces a most parsimonious reconciliation in a model without host-switches [7,8,11,12].

Phylogenetic tree reconciliation is the approach commonly used to investigate the coevolution of sets of organisms such as hosts and symbionts [4,13].

However, a huge number of most parsimonious reconciliations are possible (see *e.g.* [5]). While any biological interpretation of the underlying coevolution would require that all optimal solutions are enumerated and examined, this is humanly unfeasible without providing some sort of high level view of the situation. One approach allowing this would be to extract a small number of representatives, based on some notion of similarity between reconciliations.

To the best of our knowledge, only a few such notions have been proposed in the literature. One of them is based on the comparison of the number of each one of the four events (cospeciation, duplication, loss and host switch): two reconciliations are considered similar, and hence put in a same cluster, if they have the same number of each event, *i.e.* if they have the same event vector [2]. However, it is not difficult to find examples of very different reconciliations having the same number of each kind of event. Two of them are given in Figs. 1a and b.

In [3], the authors define some operators which enable to go from one reconciliation to another, and from this provide a similarity measure between two reconciliations that is the smallest number of operations needed to change one reconciliation into another. Unfortunately, with this approach, it can happen that reconciliations that appear very similar have a rather high distance, as shown for example by Figs. 1c and d. Moreover, the complexity of computing the similarity between reconciliations remains an open question, and there are thus no efficient algorithms for now.

In this work, we try to overcome the above problems by proposing, in Sect. 2, two equivalence relations that allow to identify many similar reconciliations with a single one, thereby substantially reducing the number of reconciliations that are enumerated.

In Sect. 3, we present some experimental results on real datasets which show that in most of the cases, these relations perform very well, especially when they are considered together. Finally, Sect. 4 proposes some future lines of research.

2   Equivalent Reconciliations

In this section we describe two equivalence relations for reconciliations. We choose to favour the intuition on which reconciliations we consider as equivalent and why, instead of overburdening the exposition with too many technical details.

2.1   Equivalence $\sim_1$

Given an optimal reconciliation $\varrho \in \mathcal{R}(H, S, \phi, C)$ and a vertex u of S such that arc (u, v) is mapped by ϱ as a host switch, *i.e.* v is mapped to a vertex $\varrho(v)$ that is incomparable with $\varrho(u)$, we have that u can be mapped by ϱ to anyone of the vertices of $path_H(\varrho(p(u)), \varrho(s(v)))$ without changing the cost of ϱ, as proved by the following result.

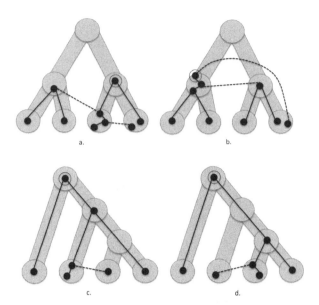

Fig. 1. a. and b. Two reconciliations with the same event vector that nevertheless are rather different. c. and d. Two reconciliations very similar with a possibly high distance (by adding arbitrarily many host vertices on the right path from the root) based on the operators. The grey tubes represent the host tree, while the black (plain or dotted) lines inside the tubes represent the symbiont tree. The roots of the symbiont trees are double lined to facilitate their recognition.

Lemma 1. *Given any two reconciliations ϱ, σ, if:*

- *there exists an arc (u, v) mapped by both ϱ and σ as a host switch, and*
- *$\varrho(w) = \sigma(w)$ for each $w \neq u$, and*
- *$\varrho(u) \neq \sigma(u)$ and $\varrho(u)$ and $\sigma(u)$ are mapped to two different vertices of $path_H(\varrho(p(u)), \varrho(s(v)))$, $\varrho(p(u))$ excluded*

then the costs associated to ϱ and σ are the same. In particular, ϱ will be optimal if and only if σ is.

The previous result leads us to consider as equivalent (using symbol $\sim_1$) all reconciliations that, for each host switch (u, v), map u on a different vertex of $path_H(\varrho(p(u)), \varrho(s(v)))$. We call the latter a *sliding path* to highlight the idea that u can be moved anywhere inside this path without modifying the cost of the reconciliation.

The following result claims an interesting property of equivalent reconciliations w.r.t. relation $\sim_1$.

Theorem 1. *Given two reconciliations $\varrho, \sigma \in \mathcal{R}(H, S, \phi, C)$, if $\varrho \sim_1 \sigma$, then they have the same event vector, i.e. $E_\varrho = E_\sigma$.*

Observe that from the previous lemma, it follows that the partition of $\mathcal{R}(H, S, \phi, C)$ induced by $\sim_1$ is finer than the partition induced by the event vector, since two reconciliations that are equivalent w.r.t. $\sim_1$ are surely equivalent w.r.t. to the event vector partition, but the opposite is not true, and this is in agreement with the fact that two reconciliations with the same event vector can be very different: in such a case, our equivalence distinguishes them.

2.2 Equivalence $\sim_2$

We now propose a second equivalence relation between optimal reconciliations. This one is motivated by the following observation. Assume there are two siblings v and w in S that are mapped by ϕ on two incomparable vertices $\phi(v)$ and $\phi(w)$ in H. If host switches are allowed, any reconciliation can equivalently map $p = p(v) = p(w)$ on a vertex that is either comparable with $\phi(v)$ and incomparable with $\phi(w)$ or vice-versa. All these solutions are equally feasible, and there is no reason to distinguish them. We can better explain this concept on the basis of the following result.

Lemma 2. *Given a reconciliation $\varrho \in \mathcal{R}(H, S, \phi, C)$, for each arc (u, v) mapped by ϱ as a host switch s.t. $\varrho(u)$ and $\varrho(p(u))$ are incomparable, $\varrho(u) = \varrho(s(v))$.*

Given optimal reconciliations in which there are two adjacent vertices u and v of S (w.l.o.g. assume $u = p(v)$) that are both associated to a host switch event, the previous result leads us to consider as equivalent (using symbol $\sim_2$) the reconciliations that map v to anyone of the vertices of H where its children are mapped. Figure 2a illustrates this concept.

More formally, we have the following:

Theorem 2. *Given any two reconciliations ϱ, σ, if in both ϱ and σ:*

- *there exists a vertex v such that the mappings of v and of one of its children, let it be w, are incomparable, while v and $s(w)$ have the same mappings, and*
- *its parent $u = p(v)$ is such that its mapping and the one of one of its children (either v or $s(v)$) are incomparable, and*
- $\varrho(t) = \sigma(t)$ *for each $t \neq v$,*

then the costs associated to ϱ and σ are the same. In particular, ϱ will be optimal if and only if σ is.

Observe that, if $u = p(v)$ is incomparable with $s(v)$ (hence we are in the context of Fig. 2b), then if the two reconciliations are optimal, if $c_l > 0$ and $c_c \leq c_d$, either the arc $(u, p(u))$ is mapped as a host switch while arc $(p(p(u)), p(u))$ is not mapped as a host switch by ϱ or σ, or there must be an ancestor of u (and thus of v), let us denote it by x, such that the following is verified:

- arc $(p(x), x)$ is mapped as a host switch by both ϱ and σ, and
- arc $(p(p(x)), p(x))$ is not mapped as a host switch by both ϱ and σ (we reach the end of the ancestry recursion), and
- all the vertices y in the path $path_S(x, u)$ are such that:
  - they are mapped to the same host vertex as v, and
  - their child that is not in the path, let us denote it by z, is such that the arc (y, z) is mapped as a host switch by both ϱ and σ.

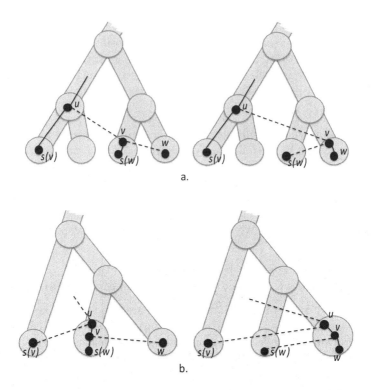

Fig. 2. The two cases in which equivalence $\sim_2$ can be applied, focusing on vertex v.

3   Results

We now show the results of some experiments performed on real datasets.

To compute the numbers of $\sim_1$ and $\sim_2$ equivalence classes, we modified the code of a well known algorithm enumerating reconciliations, *i.e.* EUCALYPT. It works by computing a matrix by means of dynamic programming, and then exploiting it to enumerate or count all reconciliations in polynomial delay. For both equivalence classes, we operated only on the first part producing a different matrix in order not to count many reconciliations falling in the same class. Our modification therefore does not affect the computational time.

As concerns the first equivalence relation, we output for each class what can be considered as a canonical representative since the produced reconciliations have some identifying properties. On the contrary, for the second equivalence, we limit ourselves to count the number of classes without enumerating them.

We selected 13 datasets which correspond to those also used in [5] and that are indicated in that paper as GL, RH, FD, COG2085, COG3715, COG4964, COG4965, PP, SFC, EC, PMP, PML, and *Wolbachia*. The latter is a dataset of our own which corresponds to arthropod hosts and a bacterium genus, *Wolbachia*, living inside the cells of their hosts. It represents a larger set (each tree has 397 leaves) than the others that were taken from the literature and where the number of leaves varies between 13 to 100. We performed the experiments using the most commonly used cost vectors, namely $(0, 1, 1, 1)$, $(0, 1, 2, 1)$, and $(0, 2, 3, 1)$ which correspond also to those presented in [5].

In all the tables, *# solutions* indicates the number of all optimal reconciliations, while $\# \sim_1$, $\# \sim_2$ and $\# \sim_2 + \sim_1$ indicate the number of equivalence classes when relations $\sim_1$, $\sim_2$ or both are applied; the last column, called *NMR*, indicates the value of the Normalized Magnitude Reduction, rounded to two digits after the decimal point, which is given by $\frac{\log(\#sol) - \log(\#\sim_1 + \sim_2)}{\log(\#sol)}$. Such value is one when all optimal solutions are reduced to a single parsimonious reconciliation when applying the two equivalences. Inversely, the closer this value is to zero, the less the two equivalences were able to reduce by similarity the number of solutions.

Observe that for *Wolbachia*, the number of solutions is so huge that, for space reason, we rounded the number to fit the table.

Table 1. Results for cost vector $(0, 1, 1, 1)$.

Dataset	# solutions	# $\sim_1$	# $\sim_2$	# $\sim_2 + \sim_1$	NMR
GL	2	2	2	2	0
RH	42	42	8	8	0.44
FD	25184	22752	224	180	0.49
COG2085	44544	36224	11	4	0.87
COG3715	1172598	777030	1888	872	0.52
COG4964	224	224	2	2	0.87
COG4965	17408	17408	4	4	0.86
PP	5120	4480	344	280	0.34
SFC	184	160	16	10	0.56
EC	16	16	13	13	0.07
PMP	2	2	1	1	1
PML	180	160	33	21	0.41
Wolbachia	$\sim 3.19 \cdot 10^{48}$	$\sim 5.72 \cdot 10^{47}$	$\sim 9.33 \cdot 10^5$	$\sim 7.68 \cdot 10^4$	0.90

We now briefly comment the results presented in Tables 1, 2 and 3.

First, note that it is not surprising that in the case of the cost vector $(0, 1, 1, 1)$, there are on average more optimal solutions than with the other cost vectors. This is due to the fact that the events that are different from cospeciation are indistinguishable in terms of cost, and this freedom of choice offers many alternatives for reaching a most parsimonious solution.

Given that both equivalence relations are primarily based on host switch mappings, we would then expect that the higher is the number of host switches, the greater would be the chance of having a lower number of equivalence classes

Table 2. Results for cost vector $(0, 1, 2, 1)$.

Dataset	# solutions	# $\sim_1$	# $\sim_2$	# $\sim_2 + \sim_1$	NMR
GL	2	2	2	2	0
RH	2208	368	1608	268	0.27
FD	408	180	48	20	0.50
COG2085	37568	3200	226	14	0.75
COG3715	9	7	4	2	0.68
COG4964	36	4	9	1	1
COG4965	640	576	4	3	0.83
PP	72	72	36	36	0.16
SFC	40	16	10	4	0.62
EC	18	18	18	18	0
PMP	2	2	1	1	1
PML	2	2	1	1	1
Wolbachia	$\sim 1.01 \cdot 10^{47}$	$\sim 3.77 \cdot 10^{44}$	$\sim 2.92 \cdot 10^{8}$	$\sim 2.42 \cdot 10^{4}$	0.91

Table 3. Results for cost vector $(0, 2, 3, 1)$.

Dataset	# solutions	# $\sim_1$	# $\sim_2$	# $\sim_2 + \sim_1$	NMR
GL	2	2	2	2	0
RH	288	48	288	48	0.32
FD	80	16	10	2	0.84
COG2085	46656	1344	540	10	0.79
COG3715	33	2	33	2	0.80
COG4964	54	6	18	2	0.83
COG4965	6528	448	94	5	0.82
PP	72	72	36	36	0.16
SFC	40	16	10	4	0.62
EC	16	16	16	16	0
PMP	18	18	10	10	0.20
PML	11	6	7	4	0.42
Wolbachia	$\sim 4.08 \cdot 10^{42}$	$\sim 1.33 \cdot 10^{36}$	$\sim 4.18 \cdot 10^{10}$	$\sim 1.15 \cdot 10^{3}$	0.93

w.r.t. the total number of solutions. Equivalence $\sim_2$ depends further on the relative position of such host switches, that is, on whether the vertices involved in a host switch are ancestors of one another, and on how long is such ancestor path in H. It is better if such paths are very long rather than if they are frequent, as there is then more chance that each long one will lead to a collapse of many solutions into a single class.

Comparing the three tables, we observe that when the cost of a host switch event is close to the cost of a loss, there is in general a smaller reduction of the number of optimal reconciliations when we pass to the $\sim_1$ equivalence classes. Intuitively, this is indeed because long sliding paths are more uncommon in this case. Inversely, the highest reductions from the total number of optimal solutions to the number of $\sim_1$ equivalence classes in general occur when the cost vectors are $(0, 1, 2, 1)$ or $(0, 2, 3, 1)$, *i.e.* when the cost of the host-switch event is higher w.r.t. the cost of a loss. In the other situations with many host switches (due either to the cost vector – *e.g.* $(0, 1, 1, 1)$ - or to the leaf-mapping, spreading close symbiont leaves to far host leaves – *e.g.* the dataset $COG2085$), equivalence $\sim_2$ performs better.

4   Perspectives

While the two equivalence relations introduced in this paper in general lead to very good results in terms of the overall goal of providing a more compact view of the solution space, we believe there are more such relations that could be explored in future.

Moreover, although the equivalence relations introduced in Sect. 2 reduce the number of enumerated reconciliations for most of the data sets, as discussed in Sect. 3, in some cases this number remains inapplicable for a direct observation of a given set of reconciliations. When less solutions are desired, it is possible to apply known clustering techniques based on a new measure of similarity between reconciliations that is able to take into account the equivalences we have defined in this paper.

More in detail, given a reconciliation $\varrho \in \mathcal{R}(H, S, \phi, C)$, preliminarily notice that each one of its host switches is univocally determined by an arc $a = (u, v)$ of S and by its mapping on a non-arc of H $(\varrho(u), \varrho(v))$, where $\varrho(u)$ and $\varrho(v)$ are incomparable. Hence, we can formalise the *host switch set of a reconciliation ϱ* as follows:

$$\Theta(\varrho) := \big\{ (u, \varrho(u), v, \varrho(v)) \subseteq V_S \times V_H \times V_S \times V_H : (u, v) \in A(S),$$

$$\varrho(u) \nsucceq_H \varrho(v) \text{ and } u \text{ is associated to a host-switch event w.r.t. } \varrho \big\}.$$

Now, given two reconciliations ϱ and σ, both in $\mathcal{R}(H, S, \phi, C)$, assume we are able to define a distance d_s between two host switches s_1 of ϱ and s_2 of σ that is able to take into account the equivalences defined in Sect. 2.

Let n_ϱ (respectively n_σ) be the number of host switches of reconciliation ϱ (respectively σ), *i.e.* $n_\varrho = |\Theta(\varrho)|$ (respectively $n_\sigma = |\Theta(\sigma)|$). We can define the complete bipartite graph $G_{\varrho,\sigma}$ on $n_\varrho + n_\sigma$ vertices, where each vertex corresponds to a host switch of either ϱ or σ (if a host switch belongs both to ϱ and σ, two vertices corresponding to it are in $G_{\varrho,\sigma}$). On $G_{\varrho,\sigma}$, an arc weight is defined: for any arc (s_ϱ, s_σ), its weight is $w(s_\varrho, s_\sigma) = d_s(s_\varrho, s_\sigma)$. Compute on $G_{\varrho,\sigma}$ a perfect matching of minimum weight $M_{\varrho,\sigma}$, with $w(M_{\varrho,\sigma}) = \sum_{e \in M_{\varrho,\sigma}} w(e)$.

We can hence define the distance between two reconciliations $\varrho, \sigma \in \mathcal{R}(H, S, \phi, C)$ as:

$$dist(\varrho, \sigma) = |n_\varrho - n_\sigma| + 2w(M_{\varrho, \sigma}).$$

This definition is able to capture similarity between reconciliations according to the two defined equivalences and seems a promising tool to define reconciliation clusters.

References

1. Bansal, M.S., Alm, E., Kellis, M.: Efficient algorithms for the reconciliation problem with gene duplication, horizontal transfer and loss. Bioinformatics **28**(12), 283–91 (2012)
2. Baudet, C., Donati, B., Sinaimeri, B., Crescenzi, P., Gautier, C., Matias, C., Sagot, M.F.: Cophylogeny reconstruction via an approximate bayesian computation. Syst. Biol. **64**(3), 416–431 (2015)
3. Chan, Y.B., Ranwez, V., Scornavacca, C.: Exploring the space of gene/species reconciliations with transfers. J. Math. Biol. **71**, 1179–1209 (2015)
4. Charleston, M.A.: Jungles: a new solution to the host/parasite phylogeny reconciliation problem. Math. Biosci. **149**(2), 191–223 (1998)
5. Donati, D., Baudet, C., Sinaimeri, B., Crescenzi, P., Sagot, M.F.: Eucalypt: efficient tree reconciliation enumerator. Algorithms Mol. Biol. **10**(1), 3 (2015)
6. Doyon, J.P., Ranwez, V., Daubin, V., Berry, V.: Models, algorithms and programs for phylogeny reconciliation. Brief Bioinform. **12**(5), 392–400 (2011)
7. Goodman, M., Czelusniak, J., Moore, G.W., Herrera, R.A., Matsuda, G.: Fitting the gene lineage into its species lineage, a parsimony strategy illustrated by cladograms constructed from globin sequences. Syst. Zool. **28**, 132–163 (1979)
8. Guigó, R., Muchnik, I., Smith, T.: Reconstruction of ancient molecular phylogeny. Mol. Phylogenet. Evol. **6**(2), 189–213 (1996)
9. Hallett, M.T., Lagergren, J.: Efficient algorithms for lateral gene transfer problems. In: Lengauer, T. (eds.) Proceedings of the Fifth Annual International Conference on Computational Biology (RECOMB 2001), pp. 149–56. ACM, New York (2001)
10. Merkle, D., Middendorf, M.: Reconstruction of the cophylogenetic history of related phylogenetic trees with divergence timing information. Theor. Biosci. **123**(4), 277–299 (2005)
11. Mirkin, B., Muchnik, I., Smith, T.: A biologically consistent model for comparing molecular phylogenies. J. Comput. Biol. **2**(4), 493–507 (1995)
12. Page, R.D.M.: Maps between trees and cladistic analysis of historical associations among genes, organisms, and areas. Syst. Biol. **43**(1), 58–77 (1994)
13. Page, R.D., Charleston, M.A.: Trees within trees: phylogeny and historical associations. Trends Ecol. Evol. **13**(9), 356–359 (1998)
14. Stolzer, M.L., Lai, H., Xu, M., Sathaye, D., Vernot, B., Durand, D.: Inferring duplications, losses, transfers and incomplete lineage sorting with nonbinary species trees. Bioinformatics **28**(18), 409–15 (2012)
15. Tofigh, A., Hallett, M., Lagergren, J.: Simultaneous identification of duplications and lateral gene transfers. IEEE/ACM Trans. Comput. Biol. Bioinf. **8**(2), 517–35 (2011)

A Quantitative and Qualitative Characterization of k-mer Based Alignment-Free Phylogeny Construction

Filippo Utro$^{(\boxtimes)}$ ⓘ, Daniel E. Platt ⓘ, and Laxmi Parida ⓘ

Computational Biology Center, IBM T. J. Watson Research,
Yorktown Heights, NY 10598, USA
{futro,watplatt,parida}@us.ibm.com

Abstract. The rapidly growing volume of genomic data, including pathogens, both invites exploration of possible phylogenetic relationships among unclassified organisms, and challenges standard techniques that require multiple sequence alignment. Further, the ability to probe variations in selection pressure e.g. among viral outbreaks, is an important characterization of the life of a virus in its biological reservoir.

In this paper, we derived the probability distribution of k-mer alignment lengths between random sequences for a given optimized score to quantify the probability that a given alignment was not better than chance, and applied it to Human Papiloma Virus (HPV), primate mtDNA, and Ebola. Even for highly variable HPV types, the number of k-mers required to significantly distinguish an alignment of related genomes from random sequences was reduced from 64 for 1-mers to 6 for 3-mers and 4 for 4-mers, indicating k-mers provide sufficient specificity to be able to characterize differences in sequences by their k-mer frequencies, allowing distances based on the k-mer frequencies to proxy for evolutionary distance. We computed mtDNA coding sequence and Ebola phylogeny construction. Primate mtDNA coding region k-mer UPGMA phylogenies reproduced most of the expected primate phylogeny. The Mantel test, applied to RAxML and Bayesian phylogenetic distances between Ebola samples versus 3-mer frequency distances, was highly significant ($\leq 1 \times 10^{-5}$). We characterized differences in selection pressure between coding and non-coding regions, and of selection in early cell cycle vs. late genes in Ebola. Coding versus non-coding regions showed evidence of purifying selection, while the early vs. late cell cycle proteins showed differences with late cycle proteins resembling influenza like immunological response, noting the g-proteins are among the late genes.

Keywords: Aligment-free · k-mers · Ebola · HPV · mtDNA

F. Utro and D.E. Platt—Contributed equally to this work.

M. Bartoletti et al. (Eds.): CIBB 2017, LNBI 10834, pp. 19–31, 2019.
https://doi.org/10.1007/978-3-030-14160-8_3

1   Introduction

In the last few years k-mers analysis has been applied to a variety of scopes in bioinformatics (e.g. [5,8,9,16,19]). Since its introduction in [2], k-mer based phylogeny construction has found interest in the promise of alignment-free phylogenies might be easily obtained [4,7,11,17] due to their computational cost and precision. Further, not only does the computational load increase dramatically with the number of taxa, but the quality of the results obtained through multiple sequence alignment algorithms drops [3]. However, it had been noted that non-alignment based phylogenetic reconstruction may suffer from some weaknesses, with an example involving primate differentiation in mtDNA [11]. This provides an opportunity to establish a baseline sensitivity for k-mer based phylogeny reconstruction. Further, a recent publication developing data from the most recent, and so far worst, Ebola outbreak [10] finds that the most recent outbreak appears to derive from an individual exposure event. Evolution within the reservoir appears to show selection pressure, while evolution in the human population reservoir following the outbreak suggesting that the pathogen is relatively benign in its non-human reservoir, allowing for evolution and selection in those hosts. This highlights the role that identification of selection pressure among genes and in outbreaks can play in understanding viral function in its reservoir and in its transmission to and among humans.

In this study, we seek to understand how k-mer specificity interacts with the changes imposed by mutations over time to yield phylogenetic reconstructability even without overt alignment. We quantify specificity in terms of the length of alignments required to exceed chance based on an extension of [12]. We argue that, starting with a common ancestor (identically aligned), each mutation differentiating lineages induces changes in the individual SNPs would each individually change the frequencies of k-mers depending on the density of mutations, generally increasing distances between vectors of k-mer frequencies. Even without specific alignments, the numbers of accumulated mutations between diverging lineages will be reflected in the differences between the k-mer frequencies. Considering that a reasonable question might be whether there is a meaningful phylogeny relating a set of hypothetically related taxa, it is worthwhile to quantify how much signal is required to resolve distances between taxa in such a phylogeny from what might be expected by chance.

We sought to understand how k-mer frequencies would capture alignment-specific information relevant to phylogenetic reconstruction. The first question is what happens to k-mer frequencies comparing aligned sequences that differ according to accumulation of mutations. The second question is the amount of specificity k-mers needed to show in order to distinguish between accumulated mutations and chance. We applied these approaches to HPV, Ebola, and primate mtDNA sequences.

2   Methods and Materials

2.1   Datasets

We downloaded mtDNA sequences for 187 human (including one Neanderthal), 187 chimpanzee (standard and pygmy), 18 baboons, 3 gibbons, 5 gorillas, and 1 orangutan from Genbank. HPV sequences identified in the International Human Papillomavirus Reference Center[1] at Karolinska Institutet were obtained from GenBank on 12/16/2014, and aligned using MAFFT [13]. We acquired 124 EBOLA segments from [10], already aligned using MAFFT.

3   Quantification of the Information Carried by k-mer

Consider two identical sequences. Distributions of the absolute frequencies of their k-mers will be identical. Introduction of a single substitution at any site will reduce the count of one k-mers, and increase the count of another one. This will produce a change of $2k$ between the two k-mer distributions in absolute (Manhattan) count. If more than one substitution occurs within k steps, so that at least one k-mer has more than one change, the impact on differences between absolute frequency distributions will be reduced from the $2k$ per mutation, since some k-mer changes capture both mutations within a single k-mer change. Insertions and deletions are similar in effect, except that the k-mers whose counts change are only those k-mers containing the deletion on the one strand compared to those on the other strand. None of the corresponding k-mers established by alignment change, so their absolute frequency distributions remain unchanged. A single deletion (or insertion) will then impact $k-1$ k-mers. Hence, it is expected that various formulations of k-mer distances would scale with genetic distance as mutations accumulate along diverging lineages, and with time to most recent common ancestor.

Given a set of alignments derived from a common ancestor, accumulated mutations, including SNPs, large deletions, or other modifications, as well as sequencing errors, can make it difficult to distinguish k-mer alignments from what might be expected by chance. To estimate how discriminating k-mer alignments are, we sought to ask how long a k-mer alignment segment would have to be in order for it to have been unlikely by random chance where nucleotides were randomly sampled from a pool matching the nucleotide frequencies of the samples.

For nucleotides $N = \{A, C, G, T\}$, k-mers will be drawn from an alphabet N^k. An alignment of two k-mers is represented as a pair of these, drawn from $\zeta \in \mathcal{S} = N^k \times N^k$. Each k-mer is associated with offsets l into putatively aligned sequences yielding $\zeta_l \in \mathcal{S}$. Given species frequencies of nucleotides N, it is possible to assign a probability $q_\zeta = P(\zeta = \zeta_l) = E\left(I(\zeta = \zeta_l)\right)$ measuring the probability that a given alignment would have been observed by chance, with each of the ζ_l being considered to be i.i.d. random variables. k-mer alignments

---

[1] http://www.hpvcenter.se/html/refclones.html.

are constructed for each of the sites along aligned segments, incrementing l by one for each step as a sliding window k long. A score s_ζ is associated with each of the possible alignments $\zeta \in \mathcal{S}$. Such a score might be derived from energies, or other criterion. A total score for aligned k-mers indexed by $l \in L$ is defined by $s = \sum_{l \in L} s_{\zeta_l}$. So for a specific sequence, this score has a definite value. In the statistical model, this is a random variable. Noting that $\sum_\zeta I(\zeta = \zeta_l) = 1$, it follows that $s = \sum_\zeta \sum_{l \in L} I(\zeta = \zeta_l) s_{\zeta_l} = \sum_\zeta \sum_{l \in L} I(\zeta = \zeta_l) s_\zeta = \sum_\zeta n_\zeta s_\zeta$, where $n_\zeta = \sum_{l \in L} I(\zeta = \zeta_l)$. In the random case, the alignment score s is a random variable. It can be proved that the n_ζ as defined above are multinomially distributed such that $p[n_\zeta] = \frac{|L|!}{\prod_{\zeta \in \mathcal{S}} n_\zeta!} \prod_{\zeta \in \mathcal{S}} q_\zeta^{n_\zeta}$. Given this, it is possible to define a distribution function $f(s||L|)ds = E\left(I(s < \sum_\zeta n_\zeta s_\zeta < s + ds)\right)$. This may be evaluated assymptotically to yield $f(s|L)ds = \frac{ds}{\sqrt{2\pi\sigma_s^2}} \exp\left(\beta s + L \ln Z\left(\beta\left(\frac{s}{L}\right)\right)\right)$ where $\sigma_s^2 = N_\zeta s_\zeta^2 - L\left(\frac{s}{L}\right)$, $Z(\beta) = \sum_{\zeta \in \mathcal{S}} q_\zeta e^{-\beta s_\zeta}$, β satisfies $s = \sum_{\zeta \in \mathcal{S}} N_\zeta s_\zeta = \frac{L}{Z(\beta)} \sum_{\zeta \in \mathcal{S}} s_\zeta q_\zeta e^{-\beta s_\zeta}$, and $N_\zeta\left(\frac{s}{L}\right) = L\frac{q_\zeta e^{-\beta s_\zeta}}{Z(\beta)}$. In this case, the N_ζ dominate the contribution from the multinomial distribution to $f(s|L)ds$. It may be shown that $\sigma_s^2 \geq 0$. The distribution of scores for a mix of alignment lengths $1 \leq l \leq L_0$ uniformly distributed as $w(l) = \frac{1}{L_0}$ for $1 \leq l \leq L_0$ and zero elsewhere, will satisfy $f(s)ds = \frac{1}{L_0} \sum_l f(s|L)ds$. This may be evaluated to be $f(s)ds = \frac{ds}{L_0}\sqrt{\frac{l_0^2}{s^2}} \exp(\beta_c s)$, where $Z(\beta_c) = 1$. This condition has two roots, one with $\beta_c = 0$ that dominates the asymptoic behavior for $s < 0$, and one with $\beta_c < 0$ for $s > 0$. Note that $\frac{l_0^2}{s^2}$ is fixed by $\beta\left(\frac{s}{l_0}\right) = \beta_c$, where l_0 is the length that dominates the sum for a given s. Lastly, a distribution of lengths $g(l|s)$ may be obtained from $g(l|s)f(s)ds = f(s|l)dsw(l)$. Then $g(l|s) = \sqrt{\frac{s^2}{2\pi l_0^2 \sigma_s^2}} \exp\left(-\frac{s^2}{2m_0^2\sigma_s^2}(l - l_0)^2\right)$.

Given these considerations, it is possible to construct tests of a hypothesis that an observed set of alignments would have emerged by chance, as opposed to the hypothesis that they were derived by a systematic process such as inheritance and mutations. To this end, we note that the $N_\zeta = Lp_\zeta = Lq_\zeta e^{-\beta s_\zeta}$ are the n_ζ that dominate the distribution of s from the multinomial distribution. So for observed proportions p_ζ, it is possible to construct scores $s_\zeta = -\frac{1}{\beta_c} \ln \frac{p_\zeta}{q_\zeta}$ that will match the observed alignment frequencies (recalling $\beta_c < 0$ in this regime). The distribution $g(l|s)$ may then be used to compute probabilities of seeing aligned lengths as long as or longer than observed given an alignment score s. If the alignments observed are significantly higher, then the signal being observed by alignment-free methods would count much higher alignment among k-mers than expected by chance. This gives a method for probing how significant the observed alignment lengths are for various k-mer lengths given observed alignment frequencies p_ζ. Note that the problem considered here is the distribution of all possible alignment scores assuming random alignments with no preferred lengths, the selection of a scoring matrix optimized to extremize for observed values, and the chances of observing a length of such observed alignments by chance. This is distinct from the problem of determining the distribution of scores that would be obtained by an alignment search algorithm that optimizes scoring matrix based scores [6].

We consider test cases for that alignment drawn from primate mtDNA, HPV, and Ebola virus. While mtDNA and Ebola show very low variability, HPV types are distinguished by at least 10% variation between sequences, with values often exceeding 30%, so serves as an extreme case. By contrast, Ebola viruses must all be within 10% to be classified as Ebola. The number of k-mers required to resolve an alignment therefore serves to measure the k-mer distance required to unambiguously resolve distances between taxa in comparison to what might be expected for a random collection of unrelated taxa.

It had been noted that, among alignment-free phylogeny construction techniques, reconstruction of phylogenies from k-mer distances was "unreliable" [11]. An example given was primate phylogeny based on mtDNA. We sought to replicate that result, computing k-mer distances on mtDNA coding regions for a number of primates.

Since phylogeny inference techniques are ultimately link to distance matrices, we sought to compare phylogenies reported among Ebola virus [10] with k-mer distance-based methods by reconstructing distance matrices from reported trees, and applying a Mantel test. This removes variability due to tree-building algorithms, but retains information about similarity between specific sequences from which phylogenies could be constructed. Selection pressure promotes conservation, which reduces variability among genes important to the biological function of the organism. We sought to establish whether constructed phylogenies would reflect selection pressure by considering the differences in phylogenies constructed from coding and non-coding Ebola virus sequences. Further, we sought to characterize the level of selection pressure that appeared between genes expressed early vs. late in the viral interaction with the cell cycle.

3.1   Distances

In this study we consider several distances to apply any alignment-free philogenies. In particular, we computed 3-mer distances using Manhattan, Euclidean, and Kullback-Leibler [15] distances. As argued above, Manhattan distances most closely correspond to the accumulation of SNPs and therefore most directly corresponds to genetic distances. Euclidean distances emphasize k-mers with the largest differences in frequency between taxa. The impact is to reduce the number of important dimensions discriminating taxa observed by clustering algorithms used to build phylogenies. Likewise, from the above considerations, Kullback-Leibler distances reflect the chances two random sequences could be so similar.

The Kullback-Leibler distance is formally defined as:

$$D_{KL} = \frac{KL(P||Q) + KL(Q||P)}{2}$$

where P and Q are the empirical probabilities of the k-mers and $KL(P||Q) = H(P,Q) - H(P)$, where $H(P,Q) = -\sum_x p_x \log_2 q_x$ is referred to as the cross entropy of P and Q, and $H(P)$ is the entropy of P. Intuitively, $KL(P\|Q)$ is a measure of the information lost when Q is used to approximate P. This emerges naturally when considering the Q's as a multinomial sampling of the P's.

Other distance scales (e.g. D_2, D_2^S, D_2^*) based on correlations [17] have been introduced and extensively studied (e.g. [4,17]). While Manhattan distances arguably most closely reflect SNP counts and genetic distances, and KL distances measure the log of the likelihood that one sequence might have been sampled by chance from the other and vice versa and are approximated by Euclidean distances for smaller deviations, D_2 and related measures also reflect Euclidean deviations at smaller genetic variances, so were not included for sake of redundancy in this analysis.

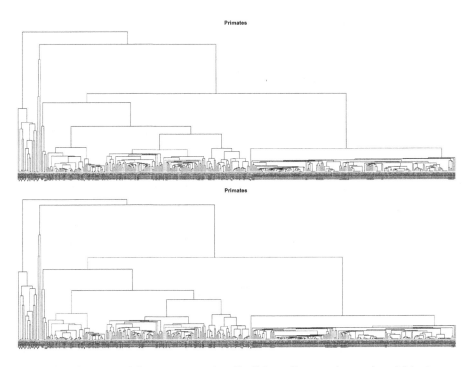

Fig. 1. Dendrogram of the primates using Euclidean distance (on top) and Manhattan distances (on the bottom). Red = Baboon (Papio papio), Yellow-green = Gibbon (Nomascus siki), Green = Gibbon (Hylobates lar), Blue-violet = Neanderthal (Homo sapiens neanderthalensis), Blue = Human (Homo sapiens), Aqua = Orangutan (Pongo pygmaeus), Orange = Gorilla (Gorilla gorilla), Violet = Chimpanzee (Pan troglodytes troglodytes), Red-violet = Bonobo (Pan paniscus) (Color figure online)

4   Results and Discussion

The characteristic lengths of k-mers required to yield a significant score are as follows. For mtDNA, the alignment length required to show two sequences align better than chance for $k = 1$ is 3.28 ± 1.75, 1.62 ± 0.86 for 2-mers, and 1.06 ± 0.55 for 3-mers. In other words, the length required to show a significant score is just one site. As k increases, k-mers are less and less likely to show alignment by chance, driving most of the s_ζ's to be more negative, with smaller probability of seeing alignment. For HPV, the length required to distinguish between alignments of random sequences from signal for 1-mer is 63.7 ± 52.4; for 2-mers, is 13.19 ± 10.76; for 3-mers, is 6.15 ± 5.15, and for 4-mers, is 3.87 ± 3.34. This shows very strikingly that, even for very diverse but clearly related sequences, the higher specificity of multimers much more strongly discriminates from alignment by chance within a much smaller length than by using monomers. Also, these lengths represent minimum alignment lengths needed to resolve taxa into a phylogeny based on alignments distinguished from chance.

For primate mtDNA clustering, we applied mean-linkage agglomerative clustering to 3-mer distances of mtDNA coding regions. Figure 1 shows that the chimpanzees cluster closer to gorillas than humans, which contradicts known phylogenies. However, the split between humans, gorillas, and chimpanzees are described as nearly polytomous. The distinction among other primates is much more clearly defined, contrary to the result previously reported for 3-mers [11].

In order to validate information comparing alignment-based phylogeny construction to k-mer based phylogeny construction, we chose to compare distance matrices rather than phylogenies derived from k-mer distances. We constructed distance matrices from the NEXUS files describing trees that [10] built for Ebola virus using a Bayesian approach and by RAxML [18]. Distances were computed as the sum along both lineages to their most recent common ancestor. We also computed 3-mer distances using Manhattan, Euclidean, and Kullback-Leibler distances. Mantel tests estimate the correlation coefficient between distance matrices, and measure significance defined in terms of 99,999 random replicates of the given rows and columns. This will therefore give a measure of whether phylogenetic trees derived from these distances similarly place taxa similarly proximal to each other without the added variable of clustering algorithms. Table 1 shows in general that all p-values were limited by the number of replicates, and highly significant. Correlations were highest for all 3-mer distances with Bayesian trees, but somewhat less than for RAxML trees. Interestingly, the RAxML trees, which showed relatively lower correlations with the 3-mer distances than the Bayesian tree distances, showed a similarly strong correlation with the Bayesian tree distances as the 3-mer distances. The 3-mer distances agreed more strongly with the Bayesian tree distances than the RAxML tree distances did. The Manhattan distances showed slightly stronger correlation with RAxML results than either Euclidean or Kullback-Leibler.

We extracted Coding DNA Sequence (CDS) and non-Coding DNA Sequence (nCDS) in the Ebola virus by selecting the sequence with fewest deletions, and using the the markers for coding and non-coding segments specified in the

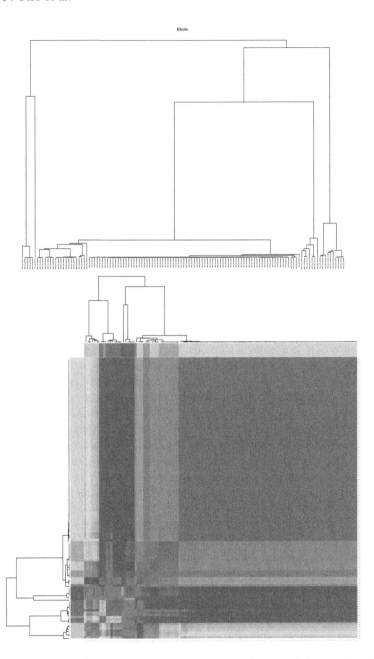

Fig. 2. Dendogram and heatmap of the strand in CDSs. Violet = Bundibugyo virus2007, Blue-violet = Tai Forest virus 1994, Blue = EBOV 1976/1977, Aqua-green = EBOV 1994-1996, Green = EBOV 2007, Yellow-Green = EOBV 2014, Orange = Reston virus 1990/1996/2008/2009, Blue = SUDV 1976, Red = SUDV 1979/2000/2004/2011/2012; Heatmap is rainbow with red = most similar, violet least similar (Color figure online)

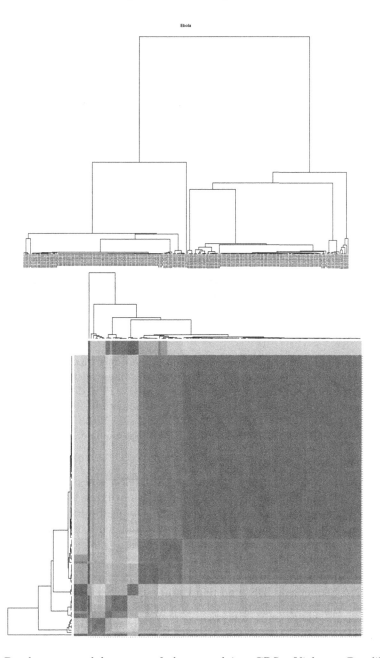

Fig. 3. Dendrogram and heatmap of the strand in nCDSs. Violet = Bundibugyo virus2007, Blue-violet = Tai Forest virus 1994, Blue = EBOV 1976/1977, Aqua-green = EBOV 1994-1996, Green = EBOV 2007, Yellow-Green = EOBV 2014, Orange = Reston virus 1990/1996/2008/2009, Blue = SUDV 1976, Red = SUDV 1979/2000/2004/2011/2012; Heatmap is rainbow with red = most similar, violet least similar (Color figure online)

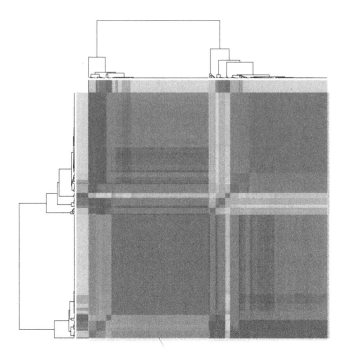

Fig. 4. Heatmap and corresponding dendogram of the k-mer distribution in strand in nCDSs and nCDSs. Violet = Bundibugyo virus2007, Blue-violet = Tai Forest virus 1994, Blue = EBOV 1976/1977, Aqua-green = EBOV 1994-1996, Green = EBOV 2007, Yellow-Green = EOBV 2014, Orange = Reston virus 1990/1996/2008/2009, Blue = SUDV 1976, Red = SUDV 1979/2000/2004/2011/2012; Heatmap is rainbow with red = most similar, violet least similar (Color figure online)

GENBANK file. We computed 3-mer distance matrices and phylogenies on these segments using complete linkage [14]. We expect purifying selection to exclude lineages with more deleterious SNPs, yielding a simple graphical identification of selection pressure. Similarly, we extracted individual genes, and compared selection pressure for genes that express early in the viral cell cycle to those later in the cycle.

Figures 2 and 3 show clustering of strands using the D_{KL} distance in the Coding DNA Sequence (CDS) and non-Coding DNA Sequence (nCDS) regions, respectively. While Fig. 4 clustered CDS and nCDS segments together. Moreover, some genes are active early in the life cycle, and others later. We characterized nucleoproteins (NP - early), RNA polymerase (L early), glycoproteins (GP - later). Therefore contrasting early versus later, we consider a combination of NP and L vs GP. Figure 5 shows the results when only these regions are taken into account. This suggests that there is much more conservation for NP and L vs GP.

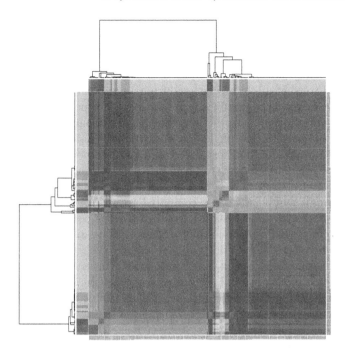

Fig. 5. Heatmap and corresponding dendogram of the k-mer distribution in strand in NP and L vs G's. Violet = Bundibugyo virus2007, Blue-violet = Tai Forest virus 1994, Blue = EBOV 1976/1977, Aqua-green = EBOV 1994-1996, Green = EBOV 2007, Yellow-Green = EOBV 2014, Orange = Reston virus 1990/1996/2008/2009, Blue = SUDV 1976, Red = SUDV 1979/2000/2004/2011/2012; Heatmap is rainbow with red = most similar, violet least similar (Color figure online)

Table 1. Mantel test with 99,999 replicates.

1st Distance matrix	2nd Distance matrix	r^2	p-value estimate
3-mer Manhattan	Bayesian tree	0.9545	$\leq 1 \times 10^{-5}$
3-mer Manhattan	RAxML tree	0.8816	$\leq 1 \times 10^{-5}$
3-mer Euclidean	Bayesian tree	0.9531	$\leq 1 \times 10^{-5}$
3-mer Euclidean	RAxML tree	0.8793	$\leq 1 \times 10^{-5}$
3-mer Kullback-Leibler	Bayesian tree	0.9578	$\leq 1 \times 10^{-5}$
3-mer Kullback-Leibler	RAxML tree	0.8746	$\leq 1 \times 10^{-5}$
Bayesian tree	RAxML tree	0.9379	$\leq 1 \times 10^{-5}$

5   Conclusion

Given the large and rapidly growing number of studies considering the relationship among multiple genomes, a rapid and simple way to extract phylogenetic relationships among putatively related taxa is increasingly important, especially where the sequences may be sufficiently phylogonetically divergent that multiple sequence alignment may be problematical. Distances determined from k-mer distributions serve as a simple way to estimate phylogenies. We have sought to quantify how much information is required to discriminate relationships among taxa compared to what might be expected from a random collection of unrelated samples. HPV serves as a great example since HPV types must show a minimum of 10% variation or more between them, and demonstrate a huge range of diversity. This makes HPV phylogeny construction challenging for alignment methods, and provides a good test case to understand how much power k-mer frequency-based distances have to resolve phylogeny-bearing mutations from random chance. In this case, we see that k-mer distances can resolve phylogenies with a very small number of bases compared to what would be expected for monomers. Therefore, k-mer distance based phylogeny construction is likely to be effective in noisy environments exploring relationships among a hypothetically related set of taxa than some prior publications have suggested.

Ebola is a fascinating virus in that it primarily evolves and survives in a wild reservoir largely out of sight until it is transmitted into the human population, at which point it is very lethal, and relatively easily transmitted. Since the reservoir allows for transmission, preserving the virus and allowing it to evolve, selection pressure preserving function of genes, and yet being pruned by immune system responses of the hosts, the virus will show typical viral evolution patterns in its hosts similar to influenza or any other virus. It is reasonable to expect that the role of g-proteins in invading cells, and the sensitivity of immune system responses to those proteins, would resemble what happens with influenza and other pathogens. However, most Ebola viruses are lethal in humans. According to [10], there is little evidence of any selection pressure once it has been transmitted to the human population. There are not enough human survivors with established immune responses selecting against older strains to show any selection pressure within an outbreak. Therefore, explorations of selection presents a probe into the evolution of Ebola within its reservoir. One possible human reservoir that may be of interest is the Reston virus, which was seen to be non-lethal. If it has not been cleared by normal immune responses, it and its descendants may be present and evolving in the human population today. The Reston virus appears to be indigenous to a reservoir in the Philippines, shows no significant impact upon transmission to humans, and immune antibodies against Reston is present in the human Philippine population [1].

Our experiment of clustering coding and non-coding 3-mer relative frequency vectors together showed that the coding and non-coding regions were more distinct from each other than the variation across strains within the regions. Further, the phylogeny of the Ebola strains were recapituated in the coding and non-coding regions. Lastly, the non-coding region showed more evidence of the survival accumulated SNPs, indicating less selection pressure against them. Accepting these as guidelines, we considered early vs. late genes, more diversity among g-proteins, perhaps indicative of pruning due to immune response within the reservoir species, similar to the phylogenies of influenza.

References

1. Barrette, R.W., et al.: Discovery of swine as a host for the reston ebolavirus. Science **325**(5937), 204–206 (2009)
2. Blaisdell, B.E.: A measure of the similarity of sets of sequences not requiring sequence alignment. Proc. Nat. Acad. Sci. **83**, 5155–5159 (1986)
3. Boyce, K., Sievers, F., Higgins, D.G.: Instability in progressive multiple sequence alignment algorithms. Algorithms Mol. Biol. **10**(1), 1–10 (2015)
4. Chan, C.X., Bernard, G., Poirion, O., Hogan, J.M., Ragan, M.A.: Inferring phylogenies of evolving sequences without muultiple sequence alignment. Sci. Rep. **4**(6504), 1–9 (2014)
5. Chor, B., Horn, D., Goldman, N., Levy, Y., Massingham, T.: Genomic DNA k-merspectra: models and modalities. Genome Biol. **10**, R108 (2009)
6. Dembo, A., Karlin, S., Zeitouni, O.: Limit distribution of maximal non-aligned two-sequence segmental score. Ann. Probab. **22**(4), 2022–2039 (1994)
7. Ferragina, P., Giancarlo, R., Greco, V., Manzini, G., Valiente, G.: Compression-based classification of biological sequences and structures via the universal similarity metric: experimental assessment. BMC Bioinform. **8**, 252 (2007)
8. Giancarlo, R., Scaturro, D., Utro, F.: Textual data compression in computational biology: a synopsis. Bioinformatics **25**, 1575–1586 (2009)
9. Giancarlo, R., Rombo, S.E., Utro, F.: Epigenomic k-mer dictionaries: shedding light on how sequence composition influences nucleosome positioning in vivo. Bioinformatics **31**, 2939–2946 (2015)
10. Gire, S.K., et al.: Genomic surveillance elucideates ebola virus origin and transmission during the 2014 outbreak. Science **345**, 1369–1372 (2014)
11. Haubold, B.: Alignment-free phylogenetics and population genetics. Briefings Bioinform. **15**, 407–418 (2013)
12. Karlin, S., Altschul, S.F.: Methods for assessing the statistical significance of molecular sequence features by using general scoring functions. PNAS **87**(6), 2264–2268 (1990)
13. Katoh, K., Standley, D.M.: Mafft multiple sequence alignment software versions 7: improvements in performance and usability. Mol. Biol. Evol. **30**(4), 772–780 (2013)
14. Kaufman, L., Rousseeuw, P.J.: Finding Groups in Data: An Introduction to Cluster Analysis. Wiley, New York (1990)
15. Kullback, S., Leibler, R.: On information and sufficiency. Ann. Math. Statist. **22**, 79–86 (1951)
16. Lo Bosco, G.: Alignment free dissimilarities for nucleosome classification. In: Angelini, C., Rancoita, P.M.V., Rovetta, S. (eds.) CIBB 2015. LNCS, vol. 9874, pp. 114–128. Springer, Cham (2016). https://doi.org/10.1007/978-3-319-44332-4_9
17. Song, K., Ren, J., Reinert, G., Deng, M., Waterman, M.S., Fengzhu, S.: New developments of alignment-free sequence comparison: measures, statistics and next-generation sequencing. Briefings Bioinform. **15**(3), 343–353 (2014)
18. Stamatakis, A.: Raxml version 8: a tool for phylogenetic analysis and post-analysis of large phylogenies. Bioinformatics **30**(9), 1312–1313 (2014)
19. Utro, F., Di Benedetto, V., Corona, D.F., Giancarlo, R.: The intrinsic combinatorial organization and information theoretic content of a sequence are correlated to the DNA encoded nucleosome organization of eukaryotic genomes. Bioinformatics **32**, 835–842 (2016)

Cancer Mutational Signatures Identification with Sparse Dictionary Learning

Veronica Tozzo[✉] and Annalisa Barla

Department of Informatics, Bioengineering, Robotics and System Engineering,
University of Genoa, 16146 Genoa, Italy
veronica.tozzo@dibris.unige.it

Abstract. Somatic DNA mutations are a characteristic of cancerous cells, being usually key in the origin and development of cancer. In the last few years, somatic mutations have been studied in order to understand which processes or conditions may generate them, with the purpose of developing prevention and treatment strategies. In this work we propose a novel sparse regularised method that aims at extracting mutational signatures from somatic mutations. We developed a pipeline that extracts the dataset from raw data and performs the analysis returning the signatures and their relative usage frequencies. A thorough comparison between our method and the state of the art procedure reveals that our pipeline can be used alternatively without losing information and possibly gaining more interpretability and precision.

Keywords: Somatic mutations · Mutational signatures ·
Dictionary learning · Sparsity

1 Scientific Background

Environmental factors as well as failings in the biological mechanisms of DNA replication and repairing are among the main causes of somatic DNA mutations, possibly leading to the pathogenesis of cancer.

In this context, one of the main challenges is to distinguish which phenomena or environmental processes generated such mutations, with the aim of better understanding each patient's history and, possibly, devising a personalized therapy. The problem translates into discovering the so-called *mutational signatures*, that are patterns of mutations characterizing the diseases. This is possible using genome sequencing techniques in combination with statistical and computational methods. A methodology for extracting mutational signatures was recently proposed by [2]. It is based on Non-negative Matrix Factorization (NMF) [5] and a variation of k-means [4] in which a one-to-one assignment based on cosine similarity is performed between clusters and signatures. In this method, from now on addressed as NMF-CL, each patient is represented with a vector of mutations

M. Bartoletti et al. (Eds.): CIBB 2017, LNBI 10834, pp. 32–41, 2019.
https://doi.org/10.1007/978-3-030-14160-8_4

counts (see Sect. 2.1) and his feature vector is approximated with a linear combination of weighted basic patterns (mutational signatures). The strong prior imposed by the non-negativity constraint in NMF is due to the data representation of choice where negative patterns of mutations have no meaning. In [1], NMF-CL analysed 10000 samples comprising 30 different cancer types extracting 22 mutational signatures. Of these signatures 11 were linked to biological or external causes.

Here, we present an alternative methodology based on dictionary learning [3,8] implemented within a cross-validation schema (DL-CV). Dictionary learning is a general decomposition technique that aims at approximating the signal with a linear combination of dictionary atoms (namely signatures) and coefficients. For the type of analysis we are dealing with dictionary learning can be specialized into NMF by imposing a non-negativity constraint. DL-CV exploits the use of sparsity on both signatures and coefficients. In fact the effect of sparsity allows for a better signal approximation and a better interpretability of the results. DL-CV is implemented within a Python library[1] [11] that can also take advantage of high-performance computing facilities as it can be distributed to an arbitrary number of processes, greatly reducing the computational time.

1.1 DNA Somatic Mutations

Deoxyribonucleic acid (DNA) encodes all the genetic background of every living organism with an alphabet of four nucleotides: adenine (A), guanine (G), cytosine (C) and thiamine (T). After conception DNA may be altered by *somatic mutations* which are changes generally caused by environmental factors (e.g., ultraviolet light). They are called *simple nucleotide variations* (SNVs) and they may consist of insertions, deletions or substitutions. They are predominantly linked to diseases, one of them being cancer.

2 Materials and Methods

The data set used for this paper is a 21 breast cancer whole genomes dataset [2] that was sequenced and pre-processed in [6]. For each sample, the considered data representation counts the number of occurrences of six main substitutions (C→T, C→G, C→A, T→C, T→A, T→G) within the context of the nucleotides immediately in 5′ and 3′ position w.r.t. the mutated base. Therefore, each sample is represented by a 96 dimensional vector counting 6 types of substitutions × 4 types of 5′ base × 4 types of 3′ base. Mutations types (data matrix columns) whose sum is lower than the 1% of the total were discarded as in [2]. The derived data matrix is $X \in \mathbb{R}_+^{21 \times 88}$.

---

[1] Code publicly available at https://github.com/slipguru/dalila under Free BDS license.

2.1   Learning Methods

Dictionary Learning. Dictionary learning is a machine learning method that approximates a data matrix $X \in \mathbb{R}^{n \times p}$ with two matrices assuming a linear combination $X \approx CD$. The output consists of a dictionary matrix $D^* \in \mathbb{R}^{k \times p}$ of basic patterns (signatures) and the coefficients matrix $C^* \in \mathbb{R}^{n \times k}$ which represents how much each signature is active in a sample. The problem is formalized in Eq. (1).

$$C^*, \ D^* \ \leftarrow \ \underset{\substack{C \in \mathcal{A} \subset \mathbb{R}^{n \times k}, \\ D \in \mathcal{B} \subset \mathbb{R}^{k \times p}}}{\operatorname{argmin}} \ \|X - CD\|_F^2 + \Phi(C) + \Psi(D) \tag{1}$$

where Φ and Ψ are penalties that impose prior knowledge of the minimization problem and $\mathcal{A}$ and $\mathcal{B}$ are constraint subsets of the real space.

To solve such problem we adopt an alternating proximal gradient descent algorithm which is appropriate for functionals with the following properties [3]: (i) all terms may be non-convex; (ii) the residual error is differentiable in one variable keeping the other fixed; (iii) the penalty terms involve only one variable at the time and they may be non-differentiable. This allows to use different types of penalties without changing the optimization flow, whose overview is given in Algorithm 1 following the theory of [3,8]. In this work we choose Φ and Ψ as Lasso penalty terms. This choice guarantees that the functional is also partially convex (i.e. it is convex in one variable keeping the other fixed).

Non-negative Matrix Factorization. NMF is a sub-class of the dictionary learning problem that assumes non-negativity for all the elements in the involved matrices. This assumption allows to develop targeted optimization methods as the one used in NMF-CL that finds its C^* and D^* as

$$C^*, \ D^* \ \leftarrow \ \underset{\substack{C \in \mathbb{R}_+^{n \times k}, \\ D \in \mathbb{R}_+^{k \times p}}}{\operatorname{argmin}} \|X - CD\|_F^2 \tag{2}$$

This method, proposed in [5], cannot be extended to penalized problems.

It is possible to use generic approaches as those cited in previous paragraph to obtain the same decomposition by imposing a non-negative constraint on the involved matrices. In fact Algorithm 1 can be specialized in NMF by applying a projection on the positive sub-space during the minimization.

Algorithm 1. Alternating proximal gradient descent

1: Random initialization of the matrices C and D
2: Let $\mathcal{F}$ be the name of $\|X - CD\|_F^2$
3: **for** $i = 0 : \text{max\_iters}$ **do**
4:      $D_{t+1} = \operatorname{prox}_{\gamma \Psi}(D_t - \gamma \nabla_D (\mathcal{F}))$
5:      $C_{t+1} = \operatorname{prox}_{\gamma \Phi}(C_t - \gamma \nabla_C (\mathcal{F}))$
6:      **if** difference between iterates $< \epsilon$ **and**
7:          different between previous and current objective function $< \epsilon$ **then**
8:              **break**

2.2 Pipelines Description

In this section we will describe in detail the pipelines used to implement DL-CV and NMF-CL. Both pipelines aim at extracting mutational signatures and the related coefficients in order to infer which signatures are the most representative of the different cancer types. The pipelines are presented in terms of functional, optimization method, robustness and parameters selection and sparsity.

NMF-CL. The main core of this pipeline is the NMF algorithm followed by a clustering step. In particular it minimize the problem in Eq. (2) through alternating multiplicative optimization [5] for non-negative matrix factorization (NMF). The robustness and the parameters selections are reached through the following steps:

1. random sampling from the dataset with Monte Carlo resampling;
2. decomposing the sampled matrix with NMF and storing the result;
3. performing clustering on the obtained signatures: the new signatures are mapped into clusters with a one-to-one mapping using cosine similarity

$$\frac{s_i s_j}{\|s_i\|_2 \|s_j\|_2}$$

   where s_i and s_j are mutational signatures;
4. the steps 1, 2 and 3 are repeated until the centroids of the clusters coincide up to a precision ϵ

The right number of signatures is selected as that number that has an high silhouette [10] value and low reconstruction error. At the end of the procedure NMF-CL requires the sparsification of the coefficients in which its coefficients are taken and normalized in order to obtain a probability distribution over the signatures. Then, the signatures that have a contribution lower than 25% are discarded (set to zero).

DL-CV. This pipeline is based on the optimisation of the problem in Eq. (1) where Φ and Ψ are ℓ_1-penalties and $\mathcal{A}$ and $\mathcal{B}$ are the positive subsets of the real space. The problem (1) is minimized via an alternating proximal gradient descent [8], for generic penalized and constrained dictionary learning. In order to find the correct number of atoms, to tune the penalization parameters and to be robust to noise in the data the dictionary learning algorithm is nested in a cross-validation procedure. Given the data matrix, for each possible combination of parameters we perform a 3-splits Montecarlo boostrap cross validation each time estimating the *Bayesian Information Criterion* (BIC).

$$\text{BIC} = -\left[\log(k) \cdot \log(n) + c \cdot \text{O}(C^*, D^*)\right] \tag{3}$$

where $\text{O}(C^*, D^*)$ is the objective function applied on the optimal matrices and c is a multiplicative constant. As the BIC is a sum of two negative terms the optimal solution will correspond to the highest possible value of the BIC. Ideally we look for the minimum number of signatures (k) that guarantee the most accurate reconstruction (*i.e.* a small value of $O(C^*, D^*)$). The assessment of the penalization terms is embedded in the evaluation of $O(C^*, D^*)$.

Pipelines Comparison. NMF-CL imposes non-negativity in the solution according to [5]. Conversely, DL-CV adopts two sparse penalisation terms which require the alternating proximal gradient descent optimization method. Moreover the solution is imposed to be non-negative. For what concerns parameters selection, DL-CV is based on cross-validation with an analysis of the *Bayesian Information Criterion* (BIC) values while NMF-CL is based on cluster stability and the analysis of the reconstruction errors. Differently from NMF-CL, where the sparsity of the solution is obtained via a thresholding procedure, in DL-CV the sparsity of the solution is enforced by the sparse penalty imposed on the coefficients matrix C^*. Finally, both pipelines perform a last step of analysis of the coefficients, *i.e.*, the computation of the frequencies for each signature obtained by counting how many samples have the related coefficients different from zero.

3   Results

In this section we present the results obtained using NMF-CL and DL-CV on the breast cancer dataset. We will first focus on how many signatures are identified by the two methods and then we will present a qualitative comparison of the patterns. Finally, we compare the two methods in terms of their approximation error.

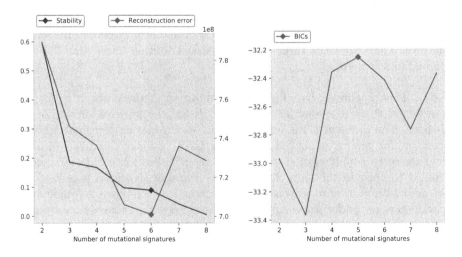

Fig. 1. Criteria used to choose the number of mutational signatures. The left panel shows the stability (mean silhouette) and the reconstruction error plotted w.r.t the number of atoms for NMF-CL. The plot on the right represents the mean BIC values for each different number of mutational signatures for DL-CV. (Color figure online)

The blue plot in Fig. 1 represents the stability criterion used to determine the right number of signatures in NMF-CL. The choice balances the trade-off between stability and reconstruction error. We favoured the reconstruction error as the stability decreases with increasing number of clusters ending up selecting 6 signatures. The red plot in Fig. 1, instead, shows the mean BIC values obtained with DL-CV. To high BIC values correspond a better model, here the highest and optimal value for BIC corresponds to a dictionary with 5 signatures.

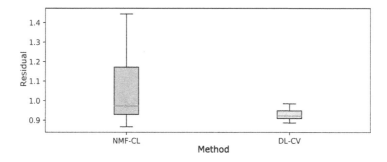

Fig. 2. Distribution of the residuals obtained for each sample with the two pipelines. The residual is computed as in Eq. (4) and it is observable that the red boxplot (DL-CV) has a lower mean and a smaller variance. (Color figure online)

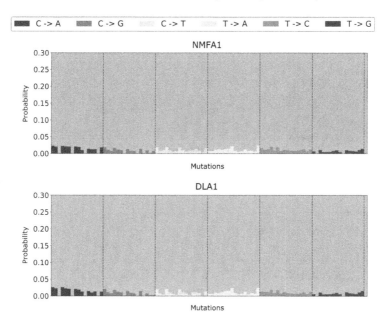

Fig. 3. This figure shows the comparison between NMFA1 and DLA1—blue and red background, respectively. Overall the two atoms share a similar shape suggesting the presence of diffused mutations across the entire DNA. (Color figure online)

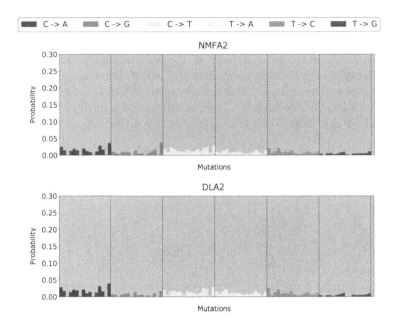

Fig. 4. This figure shows the comparison between NMFA2 and DLA2—blue and red background, respectively. Overall the two atoms share a similar shape suggesting the presence of diffused mutations across the entire DNA. (Color figure online)

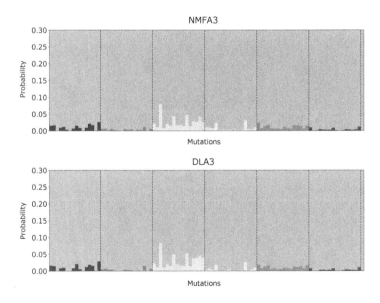

Fig. 5. This figure shows the comparison between NMFA3 and DLA3—blue and red background, respectively. Overall both methods enhance the occurrence of C→T mutation. (Color figure online)

Using the selected number of signatures we performed a qualitative comparison of the resulting signatures. The outcomes are shown in Figs. 3, 4, 5, 6, 7 where the signatures are normalized in order to represent probabilities, *i.e.* each of 96 peaks represents the probability of a specific mutation to be informative. Are comparable to the ones in [2] and they can be easily matched. Figure 4, in particular, shows three signatures that are equivalently obtained with both the procedures. Figures 6 and 7 are two cases where it is necessary to combine two signatures of one method to obtain one of the other. This behaviour ensues by the fact that dictionary learning decomposition is not unique and the

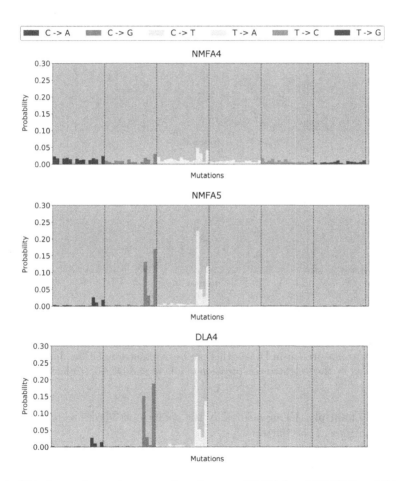

Fig. 6. This figure shows the comparison between NMFA4 and NMFA5 and DLA4—blue and red background, respectively. The main highlighted behaviour is a peak occurring in the C→T mutation. DLA4 shows a higher peak that is possibly split between NMFA4 and NMFA5. Also note that with NMF-CL method also captures background noise in the non-relevant mutations of these signatures. (Color figure online)

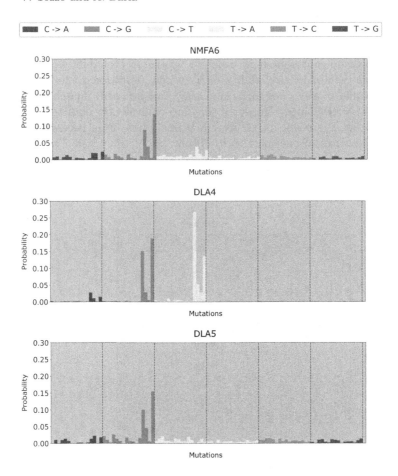

Fig. 7. This figure shows the comparison between NMFA6 and DLA4 and DLA5—blue and red background, respectively. The main highlighted behaviour is a peak occurring in the C→G mutation. While in NMFA6 there is also a relevant contribution of C→T mutation that is not present in DLA5, this can be explained by DLA4. Indeed NMFA6 can be obtained by the weighted combination of DLA4 and DLA5. (Color figure online)

possibility to multiply the signatures by the coefficients allows to recover different conformations of the patterns.

Lastly we performed an analysis on the reconstruction errors with respect to each sample. The residual for a sample i in the dataset is computed as

$$r_i = \frac{\|X_{i,:} - C^*_{i,:} D^*\|_2}{\|X_{i,:}\|_2} \tag{4}$$

where C^* and D^* are the matrices found with NMF-CL and DL-CV method. A perfect reconstruction is reached when $r_i = 0$.

All the residuals are plotted in Fig. 2 and it is noticeable that mean and variance in DL-CV are lower than in NMF-CL. This means that DL-CV achieves a better reconstruction and its errors are more similar across data points.

4 Conclusion

The proposed method is able to reproduce state of the art results and introduces new elements as sparsity and flexibility in the functional. This last novelty allows to add prior knowledge on the problem and possibly to obtain more specific results. Moreover the distributed implementation facilitates the research of the parameters by parallelizing the computation and greatly reducing the running time. In future work we plan to proceed with the coefficients analysis to assess the positive effects of post-processing elimination and to validate our methodology on new data. To this aim we will apply the pipeline on public available datasets considering also the option of analysing all the available TCGA cancer classes. Nevertheless we are aware that the use of whole-genome sequencing data may be the source of statistical issues due to the difference in the trinucleotides counts in whole genome and exomes data [9].

Acknowledgments. The authors are indebted to Dr. Alexandrov for valuable feedback on the pipeline analysis.

References

1. Alexandrov, L.B., et al.: Signatures of mutational processes in human cancer. Nature **500**(7463), 415–421 (2013)
2. Alexandrov, L.B., Nik-Zainal, S., Wedge, D.C., Campbell, P.J., Stratton, M.R.: Deciphering signatures of mutational processes operative in human cancer. Cell Rep. **3**(1), 246–359 (2013)
3. Bolte, J., Sabach, S., Teboulle, M.: Proximal alternating linearized minimization for nonconvex and nonsmooth problems. Math. Program. **146**(1), 459–494 (2014)
4. Jain, A.K.: Data clustering: 50 years beyond K-means. Pattern Recogn. Lett. **31**(8), 651–666 (2010)
5. Lee, D., Seung, H.S.: Learning the parts of objects by non-negative matrix factorization. Nature **401**(6755), 788–791 (1999)
6. Nik-Zainal, S., et al.: Mutational processes molding the genomes of 21 breast cancers. Cell **149**(5), 979–993 (2012)
7. Koboldt, D.C., et al.: VarScan: variant detection in massively parallel sequencing of individual and pooled samples. Bioinformatics **25**(17), 2283 (2009). LO
8. Rakotomamonjy, A.: Direct optimization of the dictionary learning problem. IEEE Trans. Signal Process. **61**(22), 5495–5506 (2013)
9. Ramazzotti, D., Lal, A., Liu, K., Tibshirani, R., Sidow, A.: De Novo mutational signature discovery in tumor genomes using SparseSignatures. bioRxiv. https://www.biorxiv.org/content/early/2018/08/04/384834
10. Rousseeuw, P.J.: Silhouettes: a graphical aid to the interpretation and validation of cluster analysis. J. Comput. Appl. Math. **20**, 53–65 (1987)
11. Tozzo, V., D'Amario, V., Barla, A.: Hey there's DALILA: a DictionAry LearnIng LibrAry. In: ICCSW 2017 proceedings. OpenAccess Series in Informatics (OASIcs), vol. 60, pp. 1–14 (2018)

ICING: Large-Scale Inference of Immunoglobulin Clonotypes

Federico Tomasi[1](✉) , Margherita Squillario[1] , Alessandro Verri[1] ,
Davide Bagnara[2,3] , and Annalisa Barla[1]

[1] Department of Informatics, Bioengineering, Robotics and System
Engineering (DIBRIS), Università degli studi di Genova, 16146 Genoa, Italy
federico.tomasi@dibris.unige.it
[2] Department of Experimental Medicine (DIMES), Università degli studi di Genova,
16132 Genoa, Italy
[3] The Feinstein Institute for Medical Research, North Shore-LIJ Health System,
350 Community Drive, Manhasset, NY 11030, USA

Abstract. Immunoglobulin (IG) clonotype identification is a fundamental open question in modern immunology. An accurate description of the IG repertoire is crucial to understand the variety within the immune system of an individual, potentially shedding light on the pathogenetic process. Intrinsic IG heterogeneity makes clonotype inference an extremely challenging task, both from a computational and a biological point of view. Here we present ICING, a framework that allows to reconstruct clonal families also in case of highly mutated sequences. ICING has a modular structure, and it is designed to be used with large next generation sequencing (NGS) datasets, a technology which allows the characterisation of large-scale IG repertoires. We extensively validated the framework with clustering performance metrics on the results in a simulated case. ICING is implemented in Python, and it is publicly available under FreeBSD licence at https://github.com/slipguru/icing.

Keywords: Clonotype identification · Immunoglobulin · NGS data · Cluster analysis

1 Scientific Background

The identification of immunoglobulin (IG) clonotypes is a key question in modern immunology. A clonotype is a particular combination of IGs generated by a single plasma cell clone, which is a population of cells all derived from a single progenitor cell (germline). The ability to infer clonotypes is crucial as it allows to understand how much diversity an individual has in its immune repertoire and to study immune response through B-cell clonal amplification and diversification. Indeed, understanding the variety within the immune system of an individual may potentially shed light on pathogenetic processes. In healthy individuals the

---

D. Bagnara and A. Barla—Contributed equally to this work.

M. Bartoletti et al. (Eds.): CIBB 2017, LNBI 10834, pp. 42–50, 2019.
https://doi.org/10.1007/978-3-030-14160-8_5

repertoire is expected to be extremely diverse, to guarantee the ability to respond to a wide range of antigens (*e.g.* bacteria, viruses). The diversity of the B-cell repertoire is due to the gene recombination process, where, by random selection, one for each V, D and J genes are joined together, with a simultaneous trimming and addition of random nucleotides (Fig. 1). The resulting bridging segment between V and J genes, called complementarity determining region 3 (CDR3), is the most variable and therefore important for the antigen binding [11]. Before encountering an antigen, B-cells have zero (or few) somatic mutations. Without considering mutations, the overall repertoire diversity usually comprises 10^7 to 10^8 clonotypes, with lower bounds of diversity of 10^5 and potentially as high as 10^{11} unique molecules in a single individual [4]. After the immune response, they undergo clonal amplification and somatic hypermutation, to increase the binding affinity to the antigen [8]. The potential frequency of somatic hypermutation, which can be at least 10^5–10^6 fold greater than the normal rate of mutation across the genome [9], may generate many orders of magnitude more diversity in the B-cell receptor repertoire than the 10^{11} unique molecules per individual. Therefore, intrinsic data heterogeneity makes IG clonotyping an extremely difficult task.

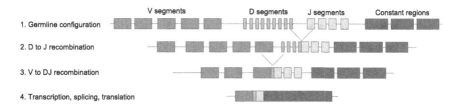

Fig. 1. IG recombination. Starting from V(D)J gene segments, one of each type is selected to produce the IG sequence. When joining two segments, some insertions and deletions (*indels*) may occur. A constant region is appended to the IG sequence after the recombination.

2   ICING

To tackle the problem of IG clonotyping inference, we developed ICING (Inferring Clonotypes of ImmuNoGlobulins), a Python library publicly available at https://github.com/slipguru/icing. The method aims at grouping IGs into clonal families, whose members derive from the same germline ancestor. Input and output data have the same format used by the Change-O suite, hence ICING is easily integrable in the usual pRESTO/Change-O pipeline [5,14]. In particular, data should be in the format produced by Change-O, that is, IGs should be represented via their V gene calls and CDR3 amminoacidic (or nucleotidic) sequence. Also, an indication of the mutation level of the sequence with respect to reference should be present, to allow for the final steps of the pipeline (Sect. 3.3).

ICING is designed to be used with a large number of data, for example coming from NGS technologies, composed of more than 10^6 sequences. The method is implemented in Python, exploiting separate processes on multi-core machines for almost each step of three sequential phases: *(i)* data shrinking, *(ii)* high-level grouping and *(iii)* fine-grained clonotype identification (Fig. 2).

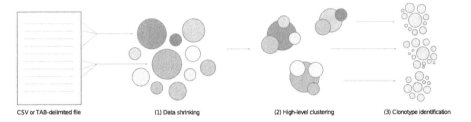

CSV or TAB-delimited file   (1) Data shrinking   (2) High-level clustering   (3) Clonotype identification

Fig. 2. ICING pipeline. Starting from a CSV or TAB-delimited file, the first step consists in grouping together sequences based on their V gene calls and CDR3 identity (data shrinking step). An high-level clustering is done on CDR3 lengths to reduce the computational workload of the third and final phase, which involves a clustering step on each of the previously found groups to obtain fine-grained IG clonotypes.

3   Materials and Methods

3.1   Synthetic Data Generation

We used *partis* [10] to generate synthetic datasets, which are characterised by an increasing number of IGs and clones, 0.05 frequency of insertions and deletions (*indels*) of maximum length of 6 nucleotides on the CDR3 sequence, and different degrees of V gene sequence mutation level. Table 1 presents an overview of the datasets.

3.2   Preprocessing

The datasets were submitted to IMGT/HighV-QUEST [1] for V(D)J genes inference, then preprocessed by a Change-O feature [5]. The outcome is a single TAB-delimited file containing the information about IGs and their metadata, such as the identification of V(D)J sequences (*i.e.*, V(D)J gene calls), V gene sequence mutation level and identification of CDR3 sequence, to be used as input to the pipeline.

3.3   Clonotype Identification

The clonotype identification step is divided into three parts.

Data Shrinking. Input data are grouped based on V gene calls (exact correspondence) and CDR3 identity (completely overlapping sequence). This allows to reduce the computational workload of next clustering steps. To each group is assigned a weight, equal to the cardinality of the group.

Table 1. Datasets overview. For reference, the total number of functional gene segments for the V/D/J regions of heavy chains in the human genome are 65/27/6 [7].

Dataset	Sequences	Clonotypes	Avg seqs/clone	Unique V genes	Unique D genes	Unique J genes	Mean (std) of V gene mutation
D1	9233	77	92.35	35	24	6	9.59 (4.64)
D2	17825	74	185.09	38	24	6	8.64 (4.46)
D3	37897	77	396.43	34	25	6	9.04 (4.51)
D4	47764	389	99.08	56	25	6	8.63 (4.30)
D5	102336	388	209.44	58	25	6	8.41 (4.70)
D6	205986	379	428.44	56	25	6	9.56 (4.46)
D7	162713	1168	109.66	58	25	6	8.72 (4.67)
D8	301978	1180	206.22	58	25	6	9.15 (4.73)
D9	589680	1185	400.26	58	25	6	8.94 (4.65)
D10	291076	2282	96.29	58	25	6	8.84 (4.46)
D11	568799	2317	187.76	58	25	6	9.12 (4.76)
D12	1208110	2358	404.30	58	25	6	9.11 (4.77)

High-Level Group Inference. This phase involves a clustering step on CDR3 lengths of previously identified groups. The outcome, which consists of high-level groups of IGs to be refined afterwards, contains IG sequences having comparable CDR3 lengths. This is done using MiniBatchKMeans clustering algorithm [12], which is computationally efficient and, more importantly, may group together very similar clusters.

Fine-Grained Group Inference. Each high-level group extracted before is then subdivided based on the actual IG distance. The distance between IGs is computed taking into account V gene calls and CDR3 sequences. In particular, the distance between two IGs is lower than infinity if and only if they have at least one V gene call in common. In such case, their actual distance is computed using a sequence distance method on their CDR3 sequences. In particular, the method implements a generic normalised distance measure based on a particular model matrix $\mathcal{M}$. Let $\|\mathcal{M}\|_{\max} = \max_{i,j} |\mathcal{M}_{ij}|$. For two sequences s and t of equal length ℓ, we defined their distance $\mathcal{D}(s,t)$ as follows:

$$\mathcal{D}(s,t) = \frac{1}{\ell \cdot \|\mathcal{M}\|_{\max}} \sum_{i=1}^{\ell} \mathcal{M}(s^i, t^i). \tag{1}$$

The choice of a specific model depends on the type of data under analysis. When $\mathcal{M} = \mathcal{H}$, where $\mathcal{H}(x,y) = 0$ if $x = y$ and 1 otherwise, the model assumes the form of a normalised Hamming distance [6].

Such distance measure allows seamless integration of different nucleotidic and amminoacidic models. ICING includes Hamming and its weighted variants, such as HS1F [16]. The models are defined between sequences of equal length. The method allows also the comparison of sequences with different lengths, by tuning a *tolerance* parameter. In such case, a standard alignment step between two sequences of different lengths may be performed before the computation of their distance, using the Smith-Waterman algorithm for sequence alignment [13].

IG sequences are characterised by an high level of mutation. Therefore, a correction function based on V gene sequence mutation level may be used to reduce distances between two IGs if mutated. This procedure encodes the uncertainty of the distance measure when dealing with highly mutated data, allowing for a more robust measure. We note that this is a step which is strongly depends on the data at hand. In our experiments, we corrected the distances between two IGs by multiplying $\mathcal{D}(s,t)$ with ν_{st}, where $\nu_{st} = 1 - \frac{m_s - m_t}{2}$, with m_s and m_t are the mutation levels of the sequences s and t, respectively.

After the design of such distance metric, fine-grained groups (*i.e.*, final clonotypes) are extracted using the DBSCAN clustering algorithm [2], which only require the parameter ϵ for the neighbourhood search of spatial distances. On top of an appropriate index structure, the algorithm can run in $O(n \log n)$ and it only needs linear memory, allowing the analysis of large-scale data.

3.4 Performance Assessment

For synthetic datasets the information about IG clonotypes is known, and it is used as ground truth. In order to evaluate clustering performance of the method, we used standard metrics such as homogeneity (HOM), completeness (COM) and V-measure (VSC), mutual information based scores, namely Adjusted Mutual Information (AMI) and Normalized Mutual Information (NMI), Adjusted Rand Index (ARI), and Fowlkes-Mallows score (FMI) [3,15]. Such measures are bound by [0,1], and no assumption is made on the cluster structure. Moreover, AMI, ARI and FMI are adjusted against chance, which is an important feature when evaluating a clustering performance in presence of a large number of clusters. Therefore, random (uniform) label assignments have scores close to 0 for measures normalised against chance.

3.5 Computing Architecture

Experiments were performed using a computing machine equipped with two Intel® Xeon® CPUs E5-2630 v3 (2.4 GHz, 8 cores each) and 128 GB of RAM[1].

4 Results

4.1 Performance Evaluation

We evaluated the method performance on the datasets shown in Table 1. In particular, Table 2 shows the clustering scores (Sect. 3.4) for datasets D1–3, obtained

---

[1] This is not representative of the amount of computational resources required by the method.

using different ICING configurations. The metric used for CDR3 sequence distance computation is the Hamming metric. The other parameters we investigated involve the neighbourhood selection radius of the DBSCAN clustering algorithm (restricted to 0.2 or 0.6), the tolerance of the difference in CDR3 sequence lengths (0, 3 or up to 6 allowed insertions or deletions), and the optional distance correction based on the V gene segment mutation level. Table 2 is ordered based on a decreasing FMI score, which, for its properties, it is the most strict of the clustering measures described in Sect. 3.4. The highest scores (close to 1) for each of the three datasets are associated to similar ICING configurations, in which the neighbourhood selection of the DBSCAN clustering algorithm is restricted to 0.2, the tolerance of the difference in sequence lengths is 0 (*i.e.*, no alignment between CDR3s needed to be done), and sequence distances are corrected based on the V gene segment mutation level. Particularly for dataset D1, the distance correction is shown to be a critical step to reliably identify IG clonotypes, as confirmed by high ARI, AMI and FMI scores (chance-corrected clustering measures). Notably, for D2 and D3 datasets, the correction gives better results when associated to a tolerance parameter of 0 or 6 nucleotides for CDR3 sequences.

The best parameters selected on datasets D1–3 were used to evaluate the results on the remaining datasets of Table 1. The results presented in Table 3

Table 2. Comparison of performance metrics between various ICING configuration on synthetic datasets. Columns are: ϵ (the DBSCAN parameter for neighbourhood selection), *tolerance* (tolerance parameter on CDR3 length), *correction* (Y for a correction based on the mutation level of V gene segments, N for no correction), followed by the clustering measures as described in Sect. 3.4. For each dataset, results are ordered by a decreasing FMI, which is the most strict of the measures for its properties.

Dataset	ϵ	Tolerance	Correction	No chance normalisation				Chance normalisation		
				HOM	COM	VSC	NMI	AMI	ARI	FMI
D1	0.2	0	Y	0.91	0.94	0.92	0.92	0.90	0.86	0.87
	0.2	6	Y	0.90	0.94	0.92	0.92	0.89	0.86	0.86
	0.2	3	Y	0.87	0.94	0.90	0.90	0.86	0.76	0.78
	0.2	6	N	0.87	0.94	0.90	0.90	0.86	0.75	0.77
	0.2	0	N	0.86	0.94	0.90	0.90	0.85	0.75	0.77
D2	0.2	0	Y	0.93	0.93	0.93	0.93	0.93	0.90	0.91
	0.2	6	Y	0.93	0.93	0.93	0.93	0.93	0.90	0.91
	0.2	3	Y	0.93	0.93	0.93	0.93	0.93	0.90	0.90
	0.2	3	N	0.92	0.93	0.92	0.92	0.91	0.88	0.88
	0.2	0	N	0.91	0.93	0.92	0.92	0.91	0.87	0.88
D3	0.2	0	Y	0.94	0.93	0.93	0.93	0.92	0.92	0.92
	0.2	3	Y	0.94	0.92	0.93	0.93	0.92	0.92	0.92
	0.2	0	N	0.92	0.93	0.92	0.92	0.91	0.89	0.89
	0.2	6	Y	0.92	0.93	0.93	0.93	0.92	0.88	0.88
	0.2	6	N	0.92	0.93	0.92	0.92	0.91	0.87	0.87

Table 3. ICING results on synthetic datasets, using the best parameters as selected in Table 2 (ϵ: 0.2, *tolerance*: 0, *correction*: Y). For each datasets, clustering measures are reported as described in Sect. 3.4.

Dataset	Sequences	No chance normalisation				Chance normalisation		
		HOM	COM	VSC	NMI	AMI	ARI	FMI
D4	47764	0.90	0.95	0.93	0.93	0.88	0.79	0.80
D5	102336	0.94	0.95	0.94	0.94	0.93	0.89	0.89
D6	205986	0.94	0.95	0.94	0.94	0.94	0.89	0.89
D7	162713	0.93	0.96	0.94	0.94	0.91	0.84	0.84
D8	301978	0.93	0.95	0.94	0.94	0.92	0.86	0.86
D9	589680	0.93	0.96	0.95	0.95	0.92	0.88	0.87
D10	291076	0.94	0.95	0.95	0.96	0.92	0.87	0.86
D11	568799	0.93	0.95	0.94	0.96	0.91	0.89	0.88
D12	1208110	0.95	0.94	0.95	0.95	0.90	0.88	0.90

show that ICING is capable to achieve high performance, which means a reliable IG sequence clonotyping, even with an increasing number of sequences. Also, the method is stable across datasets with different sizes.

4.2   Expected Clonotypes

Figure 3 shows the number of clonotypes found by ICING compared to the expected clonotypes (*ground truth*). Inferred clonotypes are very close to the

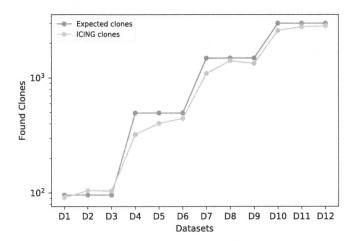

Fig. 3. Comparison between ICING clusters and expected clonotypes on synthetic datasets. For each dataset (x-axis), the number of clonotypes found by ICING is compared with the expected clonotypes (y-axis), *i.e.*, the *ground truth*. For datasets D1–3, only the best results based on FMI score (Table 2) are included.

ground truth disregarding the size of the datasets. This result, together with the high clustering performance achieved by our method (Tables 2 and 3), makes ICING a reliable framework for IG clonotype identification in real contexts, where real clonotypes are not known.

5 Conclusion

Our results show ICING to be capable of successfully identifying IG clonotypes, using synthetic data comprising highly mutated sequences, different V(D)J recombination events and *indels* on CDR3 sequences. Due to the intrinsic difficulty of validating the method on real data (where the ground truth is not known), we chose to include only the results obtained on synthetic data, where the method can be validated in relation to the ground truth.

ICING has a modular structure which allows to combine different features. In particular, the clonotype identification step has the potential to include Hamming or other arbitrary nucleotidic or amminoacidic models to compute sequence distances, arbitrary CDR3 length tolerance or V gene sequence mutation-based correction, which is an original contribution of this framework. ICING is scalable with the number of input sequences, allowing for the analysis of large-scale datasets composed of more than 10^6 sequences, which is a typical use-case when dealing with NGS data. To achieve scalability, ICING is based on a novel methodology which exploits the DBSCAN clustering algorithm, on top of an appropriate index structure. In particular, we were not able to compare our pipeline with plain Change-O which, since it is based on hierarchical clustering, has memory complexity of $O(n^2)$, thus infeasible for large datasets. However, we were able to analyse arbitrarily large datasets by exploiting all of the steps shown in Sect. 3.3, which turned out to be fundamental in our analysis.

ICING is easily integrable in the usual pRESTO/Change-O pipeline for IG analysis and it is ready to be used in real scenarios. In presence of sequences with low rate of recombination and mutation (*i.e.*, as in the case of non-healthy patients), we expect the data shrinking step (Sect. 3.3) to be highly beneficial for reducing the complexity of the algorithm, which is proportional to the number of unique CDR3 sequences and V gene calls in the dataset.

Acknowledgments. The authors would like to thank the reviewers for their helpful and constructive comments that greatly contributed to improve the final version of the paper. DB thanks *Fondazione Umberto Veronesi* for the support.

References

1. Alamyar, E., et al.: IMGT/HighV-QUEST: the IMGT® web portal for immunoglobulin (IG) or antibody and T cell receptor (TR) analysis from NGS high throughput and deep sequencing. Immunome Res. **8**(1), 26 (2012)
2. Ester, M., Kriegel, H.-P., Sander, J., Xu, X., et al.: A density-based algorithm for discovering clusters in large spatial databases with noise. In: Kdd, vol. 96, pp. 226–231 (1996)

3. Fowlkes, E.B., Mallows, C.L.: A method for comparing two hierarchical clusterings. J. Am. Stat. Assoc. **78**(383), 553–569 (1983)

4. Glanville, J., et al.: Precise determination of the diversity of a combinatorial antibody library gives insight into the human immunoglobulin repertoire. Proc. Natl. Acad. Sci. **106**(48), 20216–20221 (2009)

5. Gupta, N.T., et al.: Change-O: a toolkit for analyzing large-scale B cell immunoglobulin repertoire sequencing data. Bioinformatics **31**(20), 3356–3358 (2015)

6. Hamming, R.W.: Error detecting and error correcting codes. Bell Labs Tech. J. **29**(2), 147–160 (1950)

7. Janeway C.A., Travers, P., Walport, M., Shlomchik, M.J.: Immunobiology: The Immune System in Health and Disease, vol. 1. Current Biology Singapore (1997)

8. Kleinstein, S.H., Louzoun, Y., Shlomchik, M.J.: Estimating hypermutation rates from clonal tree data. J. Immunol. **171**(9), 4639–4649 (2003)

9. Oprea, M.L.: Antibody repertoires and pathogen recognition: the role of germline diversity and somatic hypermutation. Ph.D. thesis, Citeseer (1999)

10. Ralph, D.K., Matsen IV, F.A.: Consistency of VDJ rearrangement and substitution parameters enables accurate B cell receptor sequence annotation. PLoS Comput. Biol. **12**(1), e1004409 (2016)

11. Rock, E.P., et al.: CDR3 length in antigen-specific immune receptors. J. Exp. Med. **179**(1), 323–328 (1994)

12. Sculley, D.: Web-scale k-means clustering. In: Proceedings of the 19th International Conference on World Wide Web, pp. 1177–1178. ACM (2010)

13. Smith, T.F., Waterman, M.S.: Identification of common molecular subsequences. J. Mol. Biol. **147**(1), 195–197 (1981)

14. Vander Heiden, J.A., et al.: pRESTO: a toolkit for processing high-throughput sequencing raw reads of lymphocyte receptor repertoires. Bioinformatics **30**, 1930–1932 (2014)

15. Vinh, N.X., Epps, J., Bailey, J.: Information theoretic measures for clusterings comparison: is a correction for chance necessary? In: Proceedings of the 26th Annual International Conference on Machine Learning, pp. 1073–1080. ACM (2009)

16. Yaari, G., et al.: Models of somatic hypermutation targeting and substitution based on synonymous mutations from high-throughput immunoglobulin sequencing data. Front. Immunol. **4** (2013)

ADENINE: A HPC-Oriented Tool for Biological Data Exploration

Samuele Fiorini[✉][iD], Federico Tomasi[iD], Margherita Squillario[iD], and Annalisa Barla[iD]

Department of Informatics, Bioengineering, Robotics and System Engineering (DIBRIS), University of Genoa, 16146 Genoa, Italy
`samuele.fiorini@dibris.unige.it`

Abstract. ADENINE is a machine learning framework designed for biological data exploration and visualization. Its goal is to help bioinformaticians achieving a first and quick overview of the main structures underlying their data. This software tool encompasses state-of-the-art techniques for missing values imputing, data preprocessing, dimensionality reduction and clustering. ADENINE has a scalable architecture which seamlessly work on single workstations as well as on high-performance computing facilities. ADENINE is capable of generating publication-ready plots along with quantitative descriptions of the results. In this paper we provide an example of exploratory analysis on a publicly available gene expression data set of colorectal cancer samples. The software and its documentation are available at https://github.com/slipguru/adenine under FreeBSD license.

Keywords: Data exploration and visualization ·
Dimensionality reduction · Clustering ·
Software tool for bioinformatics · High-performance computing

1 Scientific Background

In biology, as well as in any other scientific domain, exploring and visualizing the collected measures is an insightful starting point for every data analysis process. For instance, the aim of a biomedical study can be detecting groups of patients that respond differently to a given treatment, or inferring possible molecular relationships among all, or a subset, of the measured variables. In both cases, bioinformaticians will be asked to extract meaningful information from collections of complex and high-dimensional measures, such as NGS data.

In these cases, a preliminary Exploratory Data Analysis (EDA) is not only a good practice, but also fundamental before further and deeper investigations can take place. To accomplish this task, several machine learning and data mining techniques were developed over the years. Among those, we recall the combined use of the following classes of methods: (i) missing values imputing, (ii) data

M. Bartoletti et al. (Eds.): CIBB 2017, LNBI 10834, pp. 51–59, 2019.
https://doi.org/10.1007/978-3-030-14160-8_6

Table 1. Pipeline building blocks available in ADENINE.

Step	Algorithm	Reference
Imputing	Mean	
	Median	
	Most frequent	
	k-nearest neighbors	[22]
Preprocessing	Recentering	
	Standardize	
	Normalize	
	Min-max	
Dimensionality reduction	Principal component analysis (PCA)	[11]
	Incremental PCA	[15]
	Randomized PCA	[10]
	Kernel PCA	[18]
	Isomap	[21]
	Locally linear embedding	[17]
	Spectral embedding	[13]
	Multidimensional scaling	[3]
	t-distributed stochastic neighbor embedding	[23]
Clustering	k-means	[2]
	Affinity propagation	[8]
	Mean shift	[4]
	Spectral	[20]
	Hierarchical	[9]
	DBSCAN	[7]

preprocessing, (iii) dimensionality reduction and (iv) unsupervised clustering (see Sect. 2).

In the last few years, a fair number of data exploration software and libraries were released. Such tools may be grouped in two families: GUI-based and command-line applications. Among the first group we recall *Divvy* [12], a software tool that performs dimensionality reduction and clustering on input data sets. *Divvy* is a light framework; however, its collection of C/C++ algorithm implementations does not cover common strategies such as kernel principal component analysis [18] or hierarchical clustering [9] and it does not offer strategies to perform automatic discovery of the number of clusters. The most notable project that spans between the two families is *Orange* [6], a data mining software suite that offers both visual programming front-end and Python APIs. In the context of data exploration, *Orange* can be successfully employed. However, in order to test different data analysis pipelines, each one must be manually created as it does not support their automatic generation. Moreover, large data sets are difficult to analyze with both *Divvy* and *Orange* as they can run only on a single workstation, lacking of distributed computing support.

In this paper, we present ADENINE, a command-line Python tool for biological data exploration and visualization that, starting from a set of unsupervised algorithms, creates textual and graphical reports of an arbitrary number of pipelines. Missing data imputing, preprocessing, dimensionality reduction and clustering strategies are considered as building blocks for constructing data analysis pipelines. The user is simply required to specify the input data and to select the desired blocks. ADENINE, then, takes care of generating and running the pipelines obtained by all possible combinations of the selected blocks. Every algorithm implementation of the presented software tool is inherited, or extended, from SCIKIT-LEARN [14] which is, to the best of our knowledge, the most complete machine learning open source Python library available online.

ADENINE natively supports data integration with the NCBI Gene Expression Omnibus (GEO) archive [1], which data sets can be retrieved specifying their GEO accession number.

Thanks to its scalable architecture, ADENINE pipelines can seamlessly run in parallel as separate Python processes on single workstations or MPI[1] tasks in high-performance computing (HPC) cluster facilities. This remarkable feature allows to explore and visualize massive amounts of data in a reasonable computational time. Moreover, as ADENINE makes large use of NUMPY and SCIPY, it automatically benefits from their bindings with optimized linear algebra libraries (such as OpenBLAS or Intel® MKL).

2   ADENINE

ADENINE is developed around the data analysis concept of *pipeline*. A pipeline is a sequence of the following fundamental steps: (i) missing values imputing, (ii) data preprocessing, (iii) dimensionality reduction and (iv) unsupervised clustering. For each task, different off-the-shelf algorithms are available (see Table 1).

Data collected in biomedical research studies often present missing values. Devising imputing strategies is a common practice [5] to deal with such issue. ADENINE offers an improved version of the `Imputer` class provided by SCIKIT-LEARN. In addition to the pre-existent feature-wise *mean*, *median* and *most frequent* strategies, this extension presents the k-nearest neighbors imputing method proposed for microarray data in [22].

Collecting data from heterogeneous sources may imply dealing with variables lying in very different numerical ranges and this could have a negative influence on the behavior of data analysis techniques. To tackle this issue ADENINE offers different strategies to preprocess data, such as recentering, standardizing or rescaling.

The presented software includes a set of linear and nonlinear dimensionality reduction and manifold learning algorithms that are particularly suited for

---

[1] http://mpi-forum.org/.

exploration and visualization of high-dimensional data. Such techniques rely on the fact that it is often possible to *decrease* the dimensionality of the problem estimating a low-dimensional embedding in which the data lie.

Besides offering a wide range of clustering techniques, ADENINE implements strategies and heuristics to automatically estimate parameters that yield the most suitable cluster separation. The optimal parameter selection of centroid-based algorithms follows the B-fold cross-validation strategy presented in Algorithm 1, where $\mathcal{S}(X, y)$ is the mean silhouette coefficient [16] for all input samples.

Algorithm 1. Automatic discovery of the optimal clustering parameter.

1: **for** clustering parameter k in $k_1 \ldots k_K$ **do**
2:     **for** cross-validation split b in $1 \ldots B$ **do**
3:         $X_b^{tr}, X_b^{vld} \leftarrow$ b-th training, validation set
4:         $\hat{m} \leftarrow$ fit model on X_b^{tr}
5:         $\hat{y} \leftarrow$ predict labels of X_b^{vld} according to $\hat{m}$
6:         $s_b \leftarrow$ evaluate silhouette score $\mathcal{S}(X_b^{vld}, \hat{y})$
7:     **end for**
8:     $\bar{S}_k = \frac{1}{B} \sum_{i=1}^{B} s_i$
9: **end for**
10: $k_{opt} = \arg \max_k (\bar{S}_k)$

For affinity propagation [8] and k-means [2] clustering parameters can be automatically defined (*preference* and *number of clusters*, respectively). Mean shift [4] and DBSCAN [7] offer an implicit cluster discovery. For hierarchical [9] and spectral clustering [20], no automatic discovery of clustering parameters is offered. However, graphical aids are generated to evaluate clustering performance such as dendrogram tree and eigenvalues of the Laplacian of the affinity matrix plots, respectively.

3   Gene Expression Data

In this section we show how ADENINE can be used to perform two EDAs on a gene expression microarray data set obtained from the GEO repository (accession number GSE87211). This data set was collected in a recent medical study that aimed at understanding the underlying mechanism of colorectal cancer (CRC) as well as identifying molecular biomarkers, fundamental for the disease prognostication. It is composed of 203 colorectal cancer samples and 160 matched mucosa controls. The adopted platform was the Agilent-026652 Whole Human Genome Microarray, which was used to measure the expression of 34127 probe sets.

4   Usage Example

ADENINE offers a handy tool to automatically download the data set from the GEO repository given only its accession name. It also let the user select pheno-types and/or probe sets of interest. Given these preferences, ADENINE automati-cally converts the data set from the *SOFT* format to a comma-separated values text file. To download the remote GEO data set specifying the tissue type as phenotype of interest we used the following command.

```
$ ade_GEO2csv.py GSE87211 --label_field characteristics_ch1.3.tissue
```

This automatically creates GSE87211_data.csv and GSE87211_labels.csv which contain gene expression levels and tissue type of each sample, respectively.

The first EDA aims at stratifying the samples according to their tissue type (mucosa or rectal tumor) this can be performed by executing the following com-mand.

```
$ ade_run.py ade_config.py
```

Where ade_config.py is a configuration file which should look like the snippet below.

```
1  from adenine.utils import data_source
2  data_file = 'GSE87211_data.csv'
3  labels_file = 'GSE87211_labels.csv'
4  X, y, feat_names, index = data_source.load(
5      'custom', data_file, labels_file, samples_on='rows', sep=',')
6  step1 = {'Recenter': [True], 'Normalize': [True, {'norm': ['l2']}]}
7  step2 = {'KernelPCA': [True, {'kernel': ['linear', 'rbf', 'poly']}],
8            'Isomap': [True, {'n_neighbors': 5}]}
9  step3 = {'KMeans': [True, {'n_clusters': ['auto']}]}
```

Each step variable refers to a dictionary having the name of the building block as key and a list as value. Each list has a *on\off* trigger in first position followed by a dictionary of keyword arguments for the class implementing the corresponding method. When more than one method is specified in a single step, or more than a single parameter is passed as list, ADENINE generates the pipelines composed of all possible combinations.

The configuration snippet above generates eight pipelines with similar struc-ture. The first and the second halves have recentered and ℓ_2-normalized samples, respectively. Each sample is then projected on a 2D space by isomap or by linear, Gaussian or polynomial kernel PCA. k-means clustering with automatic cluster discovery is eventually performed on each dimensionality-reduced data set, as in Algorithm 1. Results of such pipelines are all stored in a single output folder. Once this process is completed, plots and reports can be automatically generated running the following command.

```
$ ade_analysis.py results/ade_output_folder_YYYY-MM-DD_hh:mm:ss
```

The aim of the second EDA is to uncover the relationships among a set of genes known from the literature to be strongly associated with CRC. Specifically this signature is composed of the following genes: APC, KRAS, CTNNB1, TP53, MSH2, MLH1, PMS2, PTEN, SMAD4, STK11, GSK3B and AXIN2 [19]. We also considered probe sets measuring expression level of the same gene, and we labelled them with a progressive number. Three partially overlapping sublists compose this signature.

(S1) Genes fundamental for the progression of CRC (i.e. APC, KRAS, CTNNB1, TP53).

(S2) Genes relevant in the *Wnt signaling pathway*, which is strongly activated in the first phases of CRC (i.e. APC, CTNNB1, GSK3B, AXIN2).

(S3) Genes involved in hereditary syndromes which predispose to CRC (i.e. APC, MSH2, MLH1, PMS2, PTEN, SMAD4, STK11) [19].

A reduced version of the GEO data set that comprises only such genes can be easily created calling `ade_GEO2csv.py` with the option `--signature GENE_1,GENE_2,...,GENE_N`. On the same line, the option `--phenotypes P_1,P_2,...,P_M` can be used to keep only mucosa or rectal tumor samples. To run such experiment, one simply needs to select and activate the hierarchical clustering building block and to follow the same steps presented above.

For ADENINE installation instructions and for a comprehensive description of all the options available in the configuration file we refer to the online documentation and tutorials[2].

5   Results

In the first EDA, we compared the clustering performance achieved by the eight ADENINE pipelines and we reported in Fig. 1 an intuitive visualization of the results achieved by the top three, evaluated in terms of silhouette score [16]. As expected, the top performing pipelines show a clear separation between the two sample groups, as the k-means algorithm devises a domain partitioning that is consistent with the actual tissue types.

For the second EDA, the relationships among the probe sets corresponding to the genes of the signature are separately explored learning a different hierarchical clustering [9] tree for mucosa (Fig. 2a) and CRC samples (Fig. 2b), separately. The two trees are learned from different tissues, nevertheless they show some remarkable similarities. For instance, the pairs TP53-TP53.1 and MSH2-PMS2.1 always share a common parent. Interestingly, the first is a relationship between probe sets of the same gene, and the second is confirmed in literature, as MSH2 and PMS2 are both involved in hereditary non-polyposis CRC, a syndrome that predisposes for CRC. Moreover, two probe sets of the genes of *S1*, namely APC and CTNNB1, are consistently close to the root of the two trees. This suggest that the expression level of these two genes highly differs from the others.

---

[2] http://slipguru.github.io/adenine.

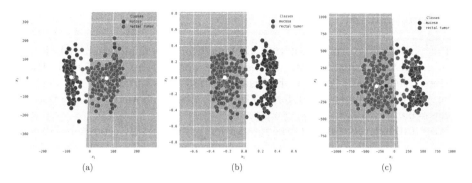

Fig. 1. Three different 2D projections of the samples of the GEO gene expression data set used in this work. Projections on the left (a), middle (b) and right (c) panes are obtained via linear PCA, Gaussian PCA and isomap, respectively. The color of each point corresponds to the actual tissue type, while the background color is automatically learned by the k-means clustering algorithm. White hexagons correspond to cluster centroids. (Color figure online)

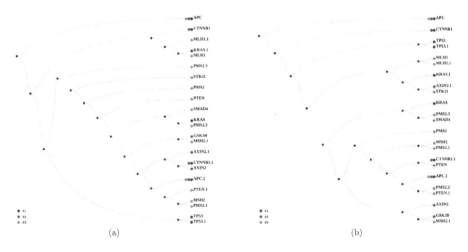

Fig. 2. An example of hierarchical trees visualization learned by two ADENINE pipelines on mucosa (a) and CRC (b) samples. Each probe set is color coded according to the corresponding sublist. This visualization provides insights on the underlying structure of the measured gene expression level.

Two interesting differences between the two trees can also be noticed. First, most of the elements of the sublist *S3*, which contains genes that enhance the risk of developing CRC, tend to be grouped together in Fig. 2b, while the same observation cannot be done for Fig. 2a. Secondly, probe sets of the genes belonging to sublists *S2* and *S3* tend more to more closely connected in Fig. 2b than in Fig. 2a.

6 Conclusions

In this paper we presented ADENINE, a biomedical data exploration and visualization tool that can seamlessly run on single workstations as well as on HPC clusters. Thanks to its scalable architecture, ADENINE is suitable for the analysis of large and high-dimensional data collections, that are nowadays ubiquitous in biomedicine.

ADENINE natively supports the integration with the GEO repository. Therefore, a user provided with the accession number of the data set of interest can select target phenotypes and genotypes and ADENINE takes care of automatically downloading the data and plugging them into the computational framework. ADENINE offers a wide range of missing values imputing, data preprocessing, dimensionality reduction and clustering techniques that can be easily selected and applied to any input data.

In this paper we showed ADENINE capabilities performing two EDAs on a CRC gene expression data set. From the obtained results we can observe that a clear discrimination between CRC and control samples can be achieved by unsupervised data analysis pipeline. Moreover, a meaningful description of the relationships among the group of genes strongly associated with CRC can be represented as hierarchical trees.

Acknowledgments. We would like to acknowledge Dr. Davide Marini for his help, assistance and support in using the high-performance computing (HPC) systems operated by the Ligurian Cluster for Marine Technologies (DLTM - http://www.dltm.it).

References

1. Barrett, T., et al.: NCBI GEO: archive for functional genomics data sets-update. Nucleic Acids Res. **41**(D1), D991–D995 (2013)
2. Bishop, C.M.: Pattern recognition. Mach. Learn. **4**, 359–422 (2006)
3. Borg, I., Groenen, P.J.F.: Modern Multidimensional Scaling: Theory and Applications. Springer, New York (2005). https://doi.org/10.1007/0-387-28981-X
4. Comaniciu, D., Meer, P.: Mean shift: a robust approach toward feature space analysis. IEEE Trans. Pattern Anal. Mach. Intell. **24**(5), 603–619 (2002)
5. De Souto, M.C.P., Jaskowiak, P.A., Costa, I.G.: Impact of missing data imputation methods on gene expression clustering and classification. BMC Bioinform. **16**(1), 64 (2015)
6. Demšar, J., et al.: Orange: data mining toolbox in Python. J. Mach. Learn. Res. **14**(1), 2349–2353 (2013)
7. Ester, M., Kriegel, H.-P., Sander, J., Xiaowei, X., et al.: A density-based algorithm for discovering clusters in large spatial databases with noise. In: KDD, vol. 96, pp. 226–231 (1996)
8. Frey, B.J., Dueck, D.: Clustering by passing messages between data points. Science **315**(5814), 972–976 (2007)
9. Friedman, J., Hastie, T., Tibshirani, R.: The Elements of Statistical Learning. Springer Series in Statistics, vol. 1. Springer, Berlin (2001). https://doi.org/10.1007/978-0-387-84858-7

10. Halko, N., Martinsson, P.-G., Tropp, J.A.: Finding structure with randomness: probabilistic algorithms for constructing approximate matrix decompositions. SIAM Rev. **53**(2), 217–288 (2011)
11. Jolliffe, I.: Principal Component Analysis. Wiley Online Library, Hoboken (2002)
12. Lewis, J.M., De Sa, V.R., Van Der Maaten, L.: Divvy: fast and intuitive exploratory data analysis. J. Mach. Learn. Res. **14**(1), 3159–3163 (2013)
13. Ng, A.Y., Jordan, M.I., Weiss, Y., et al.: On spectral clustering: analysis and an algorithm. Adv. Neural Inf. Process. Syst. **2**, 849–856 (2002)
14. Pedregosa, F., et al.: Scikit-learn: machine learning in Python. J. Mach. Learn. Res. **12**, 2825–2830 (2011)
15. Ross, D.A., Lim, J., Lin, R.-S., Yang, M.-H.: Incremental learning for robust visual tracking. Int. J. Comput. Vis. **77**(1–3), 125–141 (2008)
16. Rousseeuw, P.J.: Silhouettes: a graphical aid to the interpretation and validation of cluster analysis. J. Comput. Appl. Math. **20**, 53–65 (1987)
17. Roweis, S.T., Saul, L.K.: Nonlinear dimensionality reduction by locally linear embedding. Science **290**(5500), 2323–2326 (2000)
18. Schölkopf, B., Smola, A., Müller, K.-R.: Kernel principal component analysis. In: Gerstner, W., Germond, A., Hasler, M., Nicoud, J.-D. (eds.) ICANN 1997. LNCS, vol. 1327, pp. 583–588. Springer, Heidelberg (1997). https://doi.org/10.1007/BFb0020217
19. Schulz, W.: Molecular Biology of Human Cancers: An Advanced Student's Textbook. Springer, Dordrecht (2005). https://doi.org/10.1007/978-1-4020-3186-1
20. Shi, J., Malik, J.: Normalized cuts and image segmentation. IEEE Trans. Pattern Anal. Mach. Intell. **22**(8), 888–905 (2000)
21. Tenenbaum, J.B., De Silva, V., Langford, J.C.: A global geometric framework for nonlinear dimensionality reduction. Science **290**(5500), 2319–2323 (2000)
22. Troyanskaya, O., et al.: Missing value estimation methods for DNA microarrays. Bioinformatics **17**(6), 520–525 (2001)
23. Van der Maaten, L., Hinton, G.: Visualizing data using t-SNE. J. Mach. Learn. Res. **9**(85), 2579–2605 (2008)

Disease–Genes Must Guide Data Source Integration in the Gene Prioritization Process

Marco Frasca[1](✉)(iD), Jean Fred Fontaine[2](iD), Giorgio Valentini[1](iD),
Marco Mesiti[1](iD), Marco Notaro[1](iD), Dario Malchiodi[1](iD),
and Miguel A. Andrade-Navarro[2]

[1] Anacleto Lab – Dipartimento di Informatica, Università degli Studi di Milano,
Via Celoria 18, 20133 Milan, Italy
`{frasca,valentini,mesiti,malchiodi}@di.unimi.it, marco.notaro@unimi.it`
[2] Faculty of Biology, Johannes Gutenberg University Mainz and Institute
of Molecular Biology, Ackermannweg 4, 55128 Mainz, Germany
`{fontaine,Andrade}@uni-mainz.de`

Abstract. One of the main issues in detecting the genes involved in the etiology of genetic human diseases is the integration of different types of available functional relationships between genes. Numerous approaches exploited the complementary evidence coded in heterogeneous sources of data to prioritize disease-genes, such as functional profiles or expression quantitative trait loci, but none of them to our knowledge posed the scarcity of known disease-genes as a feature of their integration methodology. Nevertheless, in contexts where data are unbalanced, that is, where one class is largely under-represented, imbalance-unaware approaches may suffer a strong decrease in performance. We claim that imbalance-aware integration is a key requirement for boosting performance of gene prioritization (GP) methods. To support our claim, we propose an imbalance-aware integration algorithm for the GP problem, and we compare it on benchmark data with other state-of-the-art integration methodologies.

Keywords: Medical Subject Headings · Gene prioritization ·
Imbalance-aware integration · Network integration

1 Background

In the context of Network Medicine, discovering genes causing or associated with complex diseases, also known as "disease-genes", has become a central and complex challenge [2,7,16]. This process, called *gene prioritization* (GP), usually aims to supply a ranking of genes according to their involvement in the etiology of a given disease. A main issue characterizing the GP problem is the availability of a large amount of heterogeneous information about genes, ranging

M. Bartoletti et al. (Eds.): CIBB 2017, LNBI 10834, pp. 60–69, 2019.
https://doi.org/10.1007/978-3-030-14160-8_7

from protein–protein interactions to gene co-expression and functional similarity [15]. Excluding the potentially complementary evidence coming from heterogeneous data sources may be a strong limitation [3]. Several research groups have adopted computational methodologies that rely on the use of multiple heterogeneous networked-sources, and a general approach is to combine the topology of each available network into a more informative 'consensus' network, also having a larger coverage [9,18]. A common practice leverages weighted schemes to construct a linear combination of the input networks, by computing for the disease under study an informativeness coefficient for each network. For instance, in [13] the informativeness of a network has been computed as the percentage of decay in the area under the ROC curve or under the precision-recall curve of a given classifier when removing that network from the integration process. We show in this study that such a coefficient should take into account the rarity of known disease-genes characterizing most entries in existing disease ontologies, such as the Medical Subject Headings (MeSH)[1] (thousands of genetic diseases still have none or very few known causative genes). Indeed, when a disease-gene (positive gene) is rare for a given disease, it carries most information about the latter, and in principle an input source should be considered informative when it embeds information (in the form of gene connections) allowing a given classifier to correctly rank positive genes. Such an integration process is usually called "imbalance-aware", and it already led to successful results in similar contexts, such as the protein function prediction [8]. Unfortunately, the central issue represented by the rarity of disease-genes has been neglected by most existing approaches for data source integration for gene prioritization.

We argue that network integration must be imbalance-aware even for the GP problem, to improve the accuracy of gene rankings. To this purpose, we leveraged a method recently proposed for imbalance-aware integration in the context of protein function prediction, *UNIPred* (Unbalance-aware Network Integration and Prediction, [8]), and extended it in order to emphasize the important role disease-genes play in the integration process. Informally, UNIPred operates a projection of the network onto the plane, where the projected points/genes constitute the items of a new optimization problem, whose solution provides the informativeness coefficient for the input network (see [8] for theoretical details). This method has been extended by introducing a novel optimization criterion, in which the relevance to be attributed to disease-genes is associated with a free parameter, so as to easily verifying our claim. Through the network usefulness computed by UNIPred, the consensus network is built and given as input to *WGP*, a recent network-based algorithm proposed to prioritize disease-genes [6]. The overall methodology has been then validated on a benchmark data set composed of nine human networks and 708 MeSH disease terms [18].

---

[1] http://www.nlm.nih.gov/mesh/.

2 Materials

Our setup follows a benchmark proposed in [18] for data integration in the GP context. Nine human gene networks covering 8449 genes are available, considering heterogeneous data sources, as described in the following (see [18] for details about each network).

Functional interaction network – *finet*. A network covering 8441 selected proteins and containing protein–protein functional binary interactions predicted through a Naive Bayes classifier trained on a 'gold' pairwise relationships set extracted from curated pathways [19].

Human net – *hnnet*. 21 large-scale genomics and proteomics data sets from human and from orthologs in yeast, fly and worm are integrated by including distinct lines of evidence, spanning human mRNA co-expression, protein-protein interactions, protein complex, and comparative genomics data sets [10].

Cancer module network – *cmnet*. A network of 8849 genes collecting interactions derived from expression profiles in different tumors in terms of the behavior of modules of correlated genes.

Gene chemical network – *gcnet*. A network of 7649 genes constructed on the basis of direct and indirect gene–chemical interactions available at the Comparative Toxicogenomics Database (CTD) [4].

BioGRID database network – *dbnet*. BioGRID protein–protein interaction network for 8449 proteins based upon direct physical and genetic interactions constructed in [18].

BioGRID projected network – *bgnet*. An extended network from BioGRID constructed by retrieving the connection between the 8849 genes in the benchmark against all human genes in a bipartite graph, and by considering the common neighbours to determine the degree of similarity between two genes in the benchmark.

Semantic similarity networks – $\{bp, mf, cc\}net$. Three networks obtained by considering the Gene Ontology (GO, [1]) terms in the three branches annotating the considered genes: biological process (bp), molecular function (mf) and cellular component (cc). The connection between two genes is given by the maximum Resnik semantic similarity between all the terms (in that branch) the two genes are annotated with.

Gene–disease associations have been downloaded from the CTD database and include 708 selected MeSH terms having from 5 to 200 annotated disease-genes.

3 Methods

A network integration problem assumes m network sources about gene pairwise similarities are given, every source represented through a weighted undirected graph $G^{(k)} = \langle V, \boldsymbol{W}^{(k)} \rangle$, where V is the set of genes/instances (or a subset of it), $k \in \{1, 2, \ldots, m\}$ is the network index and $\boldsymbol{W}^{(k)}$ is the connection matrix: the entry $W_{i,j}^{(k)} \in [0,1]$ indicates a degree of functional similarity between genes i and j. If a data source covers just a subset of genes in V, we extended it to V by adding zeros in the corresponding entries of its connection matrix. We assume thereby in the following that all networks cover the set V. Given a disease of interest d, every gene $i \in V$ is associated with a label $y_i \in \{0, 1\}$ denoting that gene i is currently associated with d (label 1, positive gene) or not (label 0, negative gene).

The aim is to construct a composite network $G_d = \langle V, \boldsymbol{W} \rangle$ integrating all available networks, to be used to predict candidate disease-genes for d. This is performed by associating every network $G^{(k)}$ with a coefficient $r_d^{(k)}$ related to its informativeness for disease d, and then by linearly combining input networks through the obtained coefficients (see Sect. 3.2). To compute $r_d^{(k)}$ we adopt an extension of the UNIPred algorithm, briefly described in the following.

3.1 UNIPred

The UNIPred algorithm computes for every networked-source $G^{(k)}$ a relevance score taking expressly into account the disproportion between 1-labeled and 0-labeled genes for the studied genetic disease d. In particular, UNIPred operates a network projection onto the plane so that each gene $i \in V$ is associated with a labelled bi-dimensional point $P_i^{(k)}$, embedding the local imbalance in the corresponding node position. The coordinates $P_i^{(k)} \equiv (P_{i,1}^{(k)}; P_{i,2}^{(k)})$ are computed as follows:

$$P_{i,1}^{(k)} = \sum_{j \in V} W_{ij}^{(k)} \cdot y_j ,$$
$$P_{i,2}^{(k)} = \sum_{j \in V} W_{ij}^{(k)} \cdot (1 - y_j) , \tag{1}$$

In other words, $P_{i,1}^{(k)}$ is the weighted sum of 1-labeled neighbors, $P_{i,2}^{(k)}$ is the weighted sum of 0-labeled neighbors. The position of each point in the plane thereby reflects the topology of the connections towards neighboring positive and negative nodes (Fig. 1).

The algorithm then learns the straight line which best separates positive and negative points, in the sense we describe below. Since every point $i \in V$ already has a label y_i, each line separating positive and negative points is associated with the number $TP_d^{(k)}$ of positive points correctly classified (true positives), the number $FN_d^{(k)}$ of positive points wrongly classified (false negatives), and the number $FP_d^{(k)}$ of negative points wrongly classified (false positives). The optimal line is the one maximizing the F–measure: $F_d^{(k)} = \frac{2TP_d^{(k)}}{2TP_d^{(k)} + FP_d^{(k)} + FN_d^{(k)}}$. The value $\bar{F}_d^{(k)}$ corresponding to the optimal line is then considered as relevance $r_d^{(k)}$ for the input network $G^{(k)}$. The method is imbalance-aware since the F–measure by definition penalizes more heavily the misclassification of positive instances, with

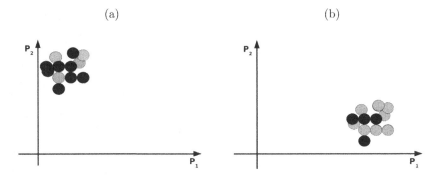

Fig. 1. Examples of distributions of points $P_i^{(k)}$ for a given network $G^{(k)}$ in which labels are unbalanced towards (a) negative points (black) and (b) positive points (light grey). In the case (a), the coordinate P_2 tends to be much larger due to the predominance of negative neighbours; as opposite, P_1 is larger in case (b), since the large majority of neighbours in average is positive.

respect to the penalty for misclassifying negatives. Moreover, maximizing $F_d^{(k)}$ moves the know labeling $\boldsymbol{y} = (y_1, \ldots, y_{|V|})$ towards a minimum of the energy of underlying Hopfield network—allowing the model to better fit the input data (see [8]).

In order to emphasize the need of attributing higher importance to positive genes, here we extend UNIPred by adopting the variant F_β of F, defined as it follows:

$$F_{\beta,d}^{(k)} := \frac{(1+\beta^2)TP_d^{(k)}}{(1+\beta^2)TP_d^{(k)} + FP_d^{(k)} + \beta^2 FN_d^{(k)}} \ . \tag{2}$$

Indeed, the parameter $\beta \in \mathbb{R}^+$ allows to regulate the importance to be assigned to the misclassification of positives rather than negatives, thus for $\beta > 1$ we assign a higher penalty to the misclassification of positives. The larger β, the more relevant are positives in determining the network coefficient $r_d^{(k)}$. Since β is dependent on the input data, we learned it through internal cross validation in our experimentations; in addition, in Sect. 4 we also supply the results of tuning β, to investigate its impact on the algorithm performance.

3.2 Constructing the Integrated Network

For a given disease of interest d, UNIPred is applied to each input network independently, obtaining the relevance vector $\boldsymbol{r}_d = (r_d^{(1)}, r_d^{(2)}, \cdots, r_d^{(m)})$. The consensus network is then constructed as a weighted sum (WS) of the corresponding adjacency matrices:

$$\boldsymbol{W} = \sum_{k=1}^{m} r_d^{(k)} \boldsymbol{W}^{(k)} \ .$$

Moreover, in order to have a baseline comparison, networks are also integrated by unweighted average sum (*US*), that is $\boldsymbol{W} = \frac{1}{m}\sum_{k=1}^{m} \boldsymbol{W}^{(k)}$.

3.3   Inferring the Gene Prioritization List

Once the consensus network $G_d = \langle V, \boldsymbol{W} \rangle$ for disease d is constructed, we are ready to face the gene prioritization problem, which is modeled as a semi-supervised ranking problem on graphs. The set of genes is assumed to be partitioned into L and U, disjoints subsets of V respectively containing the labeled and unlabeled genes, and the objective is to infer a ranking of genes in U with respect to d. Only for genes $i \in L$ the label $y_i \in \{0, 1\}$ is thereby known, and the aim is learning a function $\phi : U \to \mathbb{R}$ so as to rank higher genes susceptible to be involved in the etiology of d.

Furthermore, analogously to the integration step, the complexity of the problem is increased when the imbalance between positive and negative genes is large. Accordingly, the adopted methodology has to consider this feature of the problem to prevent a large decay of the ranking quality [5]. To learn the ranking function ϕ we employed a regression model proposed in [6], termed *WGP* (*Weighted Gene Prioritization*), able in handling the label imbalance during the prioritization process. Briefly, starting from the integrated network, WGP learns a weighted binomial regression model with log-log link function, a skewed function suitable for unbalanced data, to separate positive and negative nodes, and consequently infer the prediction for genes U using the learned regression model.

4   Results

Following the benchmark setting [18], the generalization performance of our method has been assessed through a classical 5-fold cross-validation procedure, and the results have been evaluated by using the Area Under the Receiver Operating Characteristic Curve (*AUC*) and the Precision at different Recall levels (*PxR*). In addition, we have computed the Area Under the Precision Recall Curve (*AUPRC*), to take into account the imbalance of annotated vs. unannotated genes for the MeSH disease terms. The obtained results on benchmark data show a noticeable and statistically significant improvement of validation of WGP-UNIPred algorithm with respect to the compared methods (Wilcoxon signed rank test, *p-value* < 0.01), including random walks [11], random walks with restarts, guilt-by-association methods [12] and kernelized average score functions (S_{AV} [17]). In particular S_{AV}, the top benchmark method, is based on an extension of the gene–gene similarity to non neighboring nodes by adopting a suitable kernel matrix. The score for each gene i with regard to a given disease d is defined according to a suitable distance $d(i, V_d)$ between i and the subset V_d of genes positive for d. In S_{AV}, $d(i, V_d)$ is defined as the average distance between the images in the corresponding Hilbert space of i and the elements in V_d (see [17] for details).

Figure 2 shows the overall performance, remarking both the gain of UNIPred with respect to US integration scheme and the influence of the β parameter on the performance. We only report the results of S_{AV} with weighted and unweighted sum integration, since random walk and the other compared methods achieved worse results than S_{AV}. In [18], the average AUC results across diseases have been used to weight networks according to the WS integration for S_{AV}. The β parameter has been tuned in the set of values $\{1, 2, 3, 4, 5, 10, 15, 20, 25\}$, in this first experiment, to show how it influences the model performance. To better evaluate the behaviour of our methodology, we also show results averaged across diseases with at most 10 (category '110') and more than 10 (category 'm10') associated genes. AUPRC results are not provided in the benchmark. The predictive capability of the model remarkably improves when increasing the parameter β, and more in the most unbalanced diseases (*l10*), confirming the need of imbalance-aware integration. Conversely, in US schemes, there is an almost negligible difference between *l10* and *m10* disease categories. The performance of WGP-UNIPred tends to become stable for values of β larger than 10, and, interestingly, the improvement of weighted integration is larger for WGP than for S_{AV} when compared with the corresponding unweighted strategies. This confirms that using an imbalance-aware criterion (unlike the AUC)

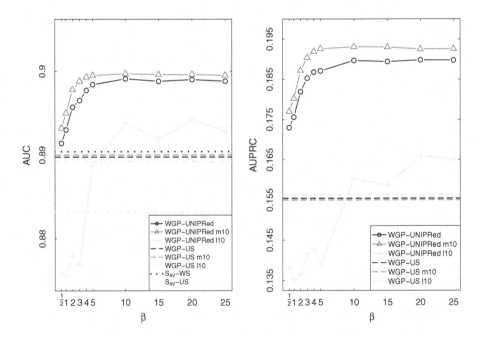

Fig. 2. Performance of WGP-UNIPRED on benchmark data. '110' and 'm10' refer to the subsets of MeSH disease terms with 5–10 and 11–200 associated genes, respectively, whereas circles correspond to results averaged across all diseases. WGP-US is the average performance across all diseases of WGP on unweighted sum data, whereas WGP-US l10 (resp. WGP-US m10) denotes the WGP performance on US data averaged across the category '110' (resp. 'm10'). S_{AV} results are averaged across all diseases.

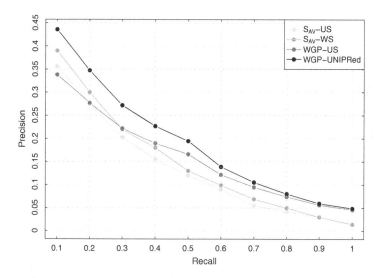

Fig. 3. PxR results achieved by the top benchmark method S_{AV} and WGP-UNIPred on both unweighted and weighted schemes.

to weight networks is more effective in this context. Apparently, the larger improvement for UNIPred compared to US scheme for *m10* with respect to *l10* terms (in both AUC and AUPRC) is quite unexpected, since *l10* terms are more unbalanced; nevertheless, since the available information for *l10* terms is very small, this behavior is likely due to overfitting phenomena. Indeed, similar works have shown that regularizing the network effectiveness for more unbalanced terms leads to better results [14]. We also compared the methods in terms of PxR (Fig. 3): in this experiment we learned β through internal cross validation. WGP-UNIPred favourably compares even in terms of PxR, outperforming S_{AV} in all experiments and WGP-US in all but 0.1 recall settings, where results are almost indistinguishable. Confirming the behaviour in terms of AUC, the UNIPred weighted sum integration led to larger improvements (mainly for lower values of recall) than the imbalance-unaware weighted integration of S_{AV}, with regard to the US corresponding results.

5   Conclusion

Experimental results supported our claim that the integration of omics data (genomics, transcriptomics, proteomics and so on) needs imbalance-aware procedures for improving the accuracy of gene prioritization lists. A state-of-the-art integration algorithm, UNIPred [8], has been used to boost the performance of a gene prioritization method, WGP [6]. By explicitly modelling the integration procedure on the exploitation of the known disease-genes, WGP-UNIPred outperformed other state-of-the-art methods in predicting gene–disease associations on public benchmark data.

Acknowledgments. This work was funded grant title *Machine learning algorithms to handle label imbalance in biomedical taxonomies*, code PSR2017_DIP_010_MFRAS, Università degli Studi di Milano.

References

1. Ashburner, M., et al.: Gene Ontology: tool for the unification of biology. The Gene Ontology Consortium. Nat. Genet. **25**(1), 25–29 (2000)
2. Barabasi, A.L., Gulbahce, N., Loscalzo, J.: Network medicine: a network-based approach to human disease. Nat. Rev. Genet. **12**(1), 56–68 (2011). https://doi.org/10.1038/nrg2918
3. Che, J., Shin, M.: A meta-analysis strategy for gene prioritization using gene expression, SNP genotype, and eQTL data. BioMed Res. Int. **2015**, 1–8 (2015). https://doi.org/10.1155/2015/576349
4. Davis, A.P., et al.: Comparative Toxicogenomics Database: a knowledgebase and discovery tool for chemical-gene-disease networks. Nucleic Acids Res. **37**(Database issue), D786–D792 (2009). https://doi.org/10.1093/nar/gkn580
5. Elkan, C.: The foundations of cost-sensitive learning. In: Proceedings of the Seventeenth International Joint Conference on Artificial Intelligence, pp. 973–978 (2001)
6. Frasca, M., Bassis, S.: Gene-disease prioritization through cost-sensitive graph-based methodologies. In: Ortuño, F., Rojas, I. (eds.) IWBBIO 2016. LNCS, vol. 9656, pp. 739–751. Springer, Cham (2016). https://doi.org/10.1007/978-3-319-31744-1_64
7. Frasca, M.: Gene2DisCo: gene to disease using disease commonalities. Artif. Intell. Med. **82**, 34–46 (2017). https://doi.org/10.1016/j.artmed.2017.08.001
8. Frasca, M., Bertoni, A., Valentini, G.: UNIPred: Unbalance-aware Network Integration and Prediction of protein functions. J. Comput. Biol. **22**(12), 1057–1074 (2015). https://doi.org/10.1089/cmb.2014.0110
9. Frasca, M., Malchiodi, D.: Exploiting negative sample selection for prioritizing candidate disease genes. Genomics Comput. Biol. **3**(3), e47 (2017). https://doi.org/10.18547/gcb.2017.vol3.iss3.e47
10. Lee, I., Blom, U.M., Wang, P.I., Shim, J.E., Marcotte, E.M.: Prioritizing candidate disease genes by network-based boosting of genome-wide association data. Genome Res. **21**(7), 1109–1121 (2011). https://doi.org/10.1101/gr.118992.110
11. Lovász, L.: Random walks on graphs: a survey. In: Miklós, D., Sós, V.T., Szőnyi, T. (eds.) Combinatorics, Paul Erdős is Eighty, vol. 2, pp. 353–398. János Bolyai Mathematical Society, Budapest (1996)
12. Marcotte, E., Pellegrini, M., Thompson, M., Yeates, T., Eisenberg, D.: A combined algorithm for genome-wide prediction of protein function. Nature **402**, 83–86 (1999)
13. Montojo, J., Zuberi, K., Shao, Q., Bader, G.D., Morris, Q.: Network assessor: an automated method for quantitative assessment of a network's potential for gene function prediction. Front. Genet. **5**, 123 (2014). https://doi.org/10.3389/fgene.2014.00123
14. Mostafavi, S., Morris, Q.: Fast integration of heterogeneous data sources for predicting gene function with limited annotation. Bioinformatics **26**(14), 1759–1765 (2010)
15. Piro, R.M., Di Cunto, F.: Computational approaches to disease-gene prediction: rationale, classification and successes. FEBS J. **279**(5), 678–696 (2012). https://doi.org/10.1111/j.1742-4658.2012.08471.x

16. Tiffin, N., Andrade-Navarro, M.A., Perez-Iratxeta, C.: Linking genes to diseases: it's all in the data. Genome Med. **1**(8), 77 (2009). https://doi.org/10.1186/gm77

17. Valentini, G., et al.: RANKS: a flexible tool for node label ranking and classification in biological networks. Bioinformatics **32**, 2872–2874 (2016). https://doi.org/10. 1093/bioinformatics/btw235

18. Valentini, G., et al.: An extensive analysis of disease-gene associations using network integration and fast kernel-based gene prioritization methods. Artif. Intell. Med. **61**(2), 63–78 (2014). https://doi.org/10.1016/j.artmed.2014.03.003

19. Wu, G., Feng, X., Stein, L.: A human functional protein interaction network and its application to cancer data analysis. Genome Biol. **11**(5), 1–23 (2010). https:// doi.org/10.1186/gb-2010-11-5-r53

Ensembling Descendant Term Classifiers to Improve Gene - Abnormal Phenotype Predictions

Marco Notaro[1] , Max Schubach[2] , Marco Frasca[1] , Marco Mesiti[1] , Peter N. Robinson[3] , and Giorgio Valentini[1(✉)]

[1] Anacleto Lab – Dipartimento di Informatica, Università degli Studi di Milano, Via Celoria 18, 20135 Milano, Italy
marco.notaro@unimi.it, {frasca,mesiti,valentini}@di.unimi.it
[2] Berlin Institute of Health (BIH),
Anna-Louisa-Karsch-Str. 2, 10178 Berlin, Germany
max.schubach@bihealth.de
[3] The Jackson Laboratory for Genomic Medicine,
10 Discovery Dr, Farmington, CT 06032, USA
Peter.Robinson@jax.org

Abstract. The Human Phenotype Ontology (HPO) provides a standard categorization of the phenotypic abnormalities encountered in human diseases and of the semantic relationship between them. Quite surprisingly the problem of the automated prediction of the association between genes and abnormal human phenotypes has been widely overlooked, even if this issue represents an important step toward the characterization of gene-disease associations, especially when no or very limited knowledge is available about the genetic etiology of the disease under study. We present a novel ensemble method able to capture the hierarchical relationships between HPO terms, and able to improve existing hierarchical ensemble algorithms by explicitly considering the predictions of the descendant terms of the ontology. In this way the algorithm exploits the information embedded in the most specific ontology terms that closely characterize the phenotypic information associated with each human gene. Genome-wide results obtained by integrating multiple sources of information show the effectiveness of the proposed approach.

Keywords: Human Phenotype Ontology ·
Hierarchical multi-label classification · Hierarchical ensemble methods ·
Gene-abnormal phenotype prediction

1 Background

The Human Phenotype Ontology (HPO) project [9] aims at providing a standard categorization of the abnormalities associated with human diseases and the semantic relationships between them. Each HPO term does not represent

© Springer Nature Switzerland AG 2019
M. Bartoletti et al. (Eds.): CIBB 2017, LNBI 10834, pp. 70–80, 2019.
https://doi.org/10.1007/978-3-030-14160-8_8

a disease, but rather it denotes individual signs or symptoms or other clinical abnormalities that characterize a disease. The HPO contains approximately 11,000 terms (still growing) and over 115,000 annotations to hereditary diseases. Moreover the HPO provides a large set of HPO annotations to approximately 4000 common diseases. The HPO is structured as a direct acyclic graph (*DAG*), where more general terms are found on the top levels of the hierarchy and the term specificity increases moving from the root to the leaves. Figure 1 shows an example of a small subset of the HPO, including all the HPO nodes that are ancestors of the *Tryptophanuria* term. In this example *Tryptophanuria* is the most specific HPO term, its parent term *Aminoaciduria* is less specific, and following the path toward the root term we find more general terms, such as *Abnormality of the urinary system*, till to the root term *Phenotypic abnormality*.

Each HPO term belongs to one of the following five subontologies: *Phenotypic abnormality, Clinical modifier, Mortality/Aging, Mode of inheritance* or *Frequency*. All the HPO relationships are *is-a* (class-subclass relationships) and are governed by the *true-path-rule* (also known as *annotation propagation rule*) [2] that can be summarized as follow: an annotation for a functional term is

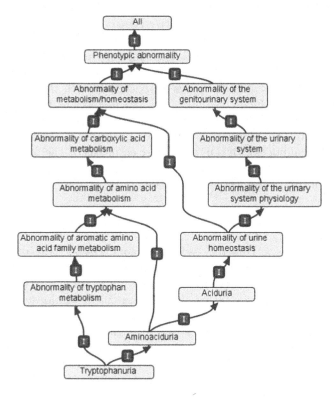

Fig. 1. Ancestor view of the HPO terms Tryptophanuria (*Phenotypic abnormality* subontology). Figure created by using OBO-Edit, an open source ontology editor.

transferred in a recursive way to its ancestors, whereas if a gene is unannotated for a class, it cannot be annotated with its descendants.

While the problem of the prediction of gene–disease associations has been widely investigated [10], the related problem of gene–phenotypic feature (i.e. HPO term) association has been largely overlooked, despite the quickly growing application of the HPO to relevant medical problems [16,22]. In principle in the contest of gene–abnormal phenotype prediction, any "flat" method that predicts labels independently of each other can be applied [21], but it may introduce significant inconsistencies in the classification due to the violation of the *true path rule* that governs the HPO taxonomy. Besides inconsistency, flat methods may also loose important a priori knowledge about the constraints of the hierarchical labeling that could enhance the accuracy of the predictions.

To overcome these limitations we recently proposed an ensemble method (*Hierarchical True path Rule for Directed Acyclic Graph* - TPR-DAG) [11, 13] that explicitly takes into account the hierarchical relationships between HPO terms, and in [11] we showed that TPR-DAG achieves competitive results with respect to state-of-the-art methods for HPO term prediction. More in general ensemble methods have been successfully applied to several branches of bioinformatics, ranging from genetic associations studies to pathogenic genetic variant prediction [7,15]. In this paper we propose a variant of the TPR-DAG algorithm, that we named *DEScendant Classifier ENSemble* (DESCENS). The novelty of DESCENS with respect to TPR-DAG consists in strongly considering the contribution of all the descendants of each node instead of only that of its children, since with the TPR-DAG algorithm the contribution of the descendants of a given node decays exponentially with their distance from the node itself, thus reducing the impact of the predictions made at the most specific levels of the ontology [17]. On the contrary DESCENS predictions are more influenced by the information embedded in the most specific terms of the taxonomy (e.g. leaf nodes), thus putting more emphasis on the terms that most characterize the gene under study, and that are those usually most informative and meaningful from a bio-medical standpoint.

2   Materials and Methods

Let $G = \langle V, E \rangle$ a Directed Acyclic Graph (DAG) with vertices $V = \{1, 2, \ldots, |V|\}$ and edges $e = (i, j) \in E, i, j \in V$. G represents the HPO taxonomy structured as a DAG, whose nodes $i \in V$ represent classes (terms) of the ontology and a directed edge $(i, j) \in E$ the hierarchical relationships between i (parent term) and j (child term). A "continuous flat multi-label scoring" predictor $f : X \rightarrow [0, 1]$ provides a score $\hat{y}_i \in [0, 1]$ that can be interpreted as the likelihood or probability that a given gene belongs to a given node/HPO term $i \in V$ of the DAG G. The set of $|V|$ flat classifiers provides a multi-label score $\hat{y} \in [0, 1]^{|V|}$: $\hat{y} = \langle \hat{y}_1, \hat{y}_2, \ldots, \hat{y}_{|V|} \rangle$. We say that a multi-label scoring y is consistent if it obeys the *true path rule*:

$$y \text{ is consistent} \iff \forall i \in V, j \in parents(i) \Rightarrow y_j \geq y_i \tag{1}$$

According to this rule the score of a parent or an ancestor node must be larger or equal than that of its children or descendants nodes.

To process and provide flat scores of the proposed hierarchical ensemble methods we used both a semi-supervised network-based approach (*RANKS* [18]) and a supervised machine learning method (Support Vector Machine – *SVM*). In our experiments we applied *RANKS* with the *average score function* and the *random walk kernel* at 1, 2 and 3 steps, i.e. kernels able to evaluate the direct neighbors and those far away 2 and 3 steps from each gene in the network. It is worth noting that *RANKS* returns a score and not a probability [12]. To make the scores comparable across classes we normalized the scores in the sense of the maximum (i.e. we divided the score values of each class by the maximum score of that class) or according to the quantile normalization [3].

After the learning phase the "flat" predictions are modified by the DESCENS algorithm, whose high-level pseudo-code is shown in Fig. 2.

```
Input:
- G =< V, E >
- V = {1, 2, ..., |V|}
- ŷ =< ŷ₁, ŷ₂, ..., ŷ|V| >,    ŷᵢ ∈ [0, 1]
begin algorithm
01:     A. dist := ∀i ∈ V ComputeMaxDist (G, root(G))
02:     B. Per-level bottom-up visit of G:
03:         for each d from max(dist) to 0 do
04:             N_d := {i|dist_i = d}
05:             for each i ∈ N_d do
06:                 Δ_i := {j ∈ desc(i)|ȳ_j > ŷ_i}
07:                 ȳ_i := (1/(1+|Δ_i|))(ŷ_i + Σ_{j∈Δ_i} ȳ_j)
08:             end for
09:         end for
10:     C. Per-level top-down visit of G:
11:         ŷ := ȳ
12:         for each d from 1 to max(dist) do
13:             N_d := {i|dist_i = d}
14:             for each i ∈ N_d do
15:                 x := min_{j∈parents(i)} ȳ_j
16:                 if (x < ŷ_i)
17:                     ȳ_i := x
18:                 else
19:                     ȳ_i := ŷ_i
20:             end for
21:         end for
end algorithm
Output:
- ȳ =< ȳ₁, ȳ₂, ..., ȳ|V| >
```

Fig. 2. DEScendant Classifier ENSemble for DAGs (DESCENS)

The block A of the algorithm (row 1) computes the maximum distance of each node from the root. To this end a method based on the Topological Sorting algorithm can be applied [5]. The block B computes a per-level bottom-up visit of the graph G (rows 2 to 9) to propagate the "positive" predictions across the hierarchy. More precisely, according to the true path rule, only the "positive" descendants of a certain node i (e.g. descendant nodes having scores larger than that of their ancestor node i) influence the prediction for the node i itself (row 6 of Fig. 2). In this way all the "positive" descendants of node i provide the same contribution to the ensemble prediction $\bar{y}_i$, by modifying the flat predictions $\hat{y}_i$.

1. *Threshold Free (TF) Strategy.* We choose as "positive" descendants those nodes that achieve a score higher than that of their ancestor node i:

$$\Delta_i := \{j \in desc(i) | \bar{y}_j > \hat{y}_i\} \tag{2}$$

   This strategy leads to the DESCENS-TF algorithm (Fig. 2).

2. *Adaptive Threshold (T) Strategy.* The threshold is selected to maximize some performance metric $\mathcal{M}(j,t)$ (e.g. F-score or *AUPRC*) estimated on the training data for the class j with respect to the threshold t. The corresponding set of positives $\forall i \in V$ is:

$$\Delta_i := \{j \in desc(i) | \bar{y}_j > t_j^*, t_j^* = \arg\max_t \mathcal{M}(j,t)\} \tag{3}$$

   For instance t_j^* can be selected from a set of $t \in (0,1)$ through internal cross-validation techniques. This strategy leads to the DESCENS-T algorithm, simply by changing row 6 in Fig. 2 with Eq. 3.

Moreover, by changing the line 7 of the algorithm in Fig. 2, we can design the "weighted" version of the DESCENS algorithm (DESCENS-W) merely adding a weight $w \in [0,1]$ to balance the contribution between the node i and that of its "positive" descendants:

$$\bar{y}_i := w\hat{y}_i + \frac{(1-w)}{|\Delta_i|} \sum_{j \in \Delta_i} \bar{y}_j \tag{4}$$

Another variant of DESCENS (named DESCENS-τ) balances the contribution between the "positive" children of a node i and that of its "positive" descendants excluding its children by adding a weight $\tau \in [0,1]$:

$$\bar{y}_i := \frac{\tau}{1 + |\phi_i|}(\hat{y}_i + \sum_{j \in \phi_i} \bar{y}_j) + \frac{1 - \tau}{1 + |\delta_i|}(\hat{y}_i + \sum_{j \in \delta_i} \bar{y}_j) \tag{5}$$

where ϕ_i are the "positive" children of i and $\delta_i = \Delta_i \setminus \phi_i$ the descendants of i without its children. If $\tau = 1$ we consider only the contribution of the "positive" children of i, and if $\tau = 0$ only the descendants that are not children contribute to the score, while for intermediate values of τ we can balance the contribution of ϕ_i and δ_i positive nodes.

Independently of which variants of the DESCENS algorithm we decide to use, "positive" predictions are "bottom-up" recursively propagated from the parents towards the ancestors of each node. The bottom-up step does not assure the consistency of the predictions. Therefore, this is guaranteed by the block C of the algorithm (row 10 to 21), where the nodes are top-down processed by level in an increasing order (from the least to the most specific terms) and the "bottom-up" scores computed at the block B are hierarchically corrected to $\bar{y}$ according to the following simple rule:

$$
\bar{y}_i := \begin{cases} \hat{y}_i & \text{if } i \in root(G) \\ \min_{j \in parents(i)} \bar{y}_j & \text{if } \min_{j \in parents(i)} \bar{y}_j < \hat{y}_i \\ \hat{y}_i & \text{otherwise} \end{cases} \tag{6}
$$

The aim of the top-down step consists in propagating the "negative" predictions towards the children and in a recursive way towards the descendants of each node. Considering the sparseness of the HPO, it is easy to see that the overall computational complexity of DESCENS algorithm is $\mathcal{O}(|V|)$.

3   Results

We downloaded physical and genetic experimental interactions relative to 4970 proteins from BioGRID (v. 3.2.106, [4]) and the integrated protein-protein interaction and functional association data for 18,172 human proteins from STRING (v. 9.1, [6]). Moreover, starting from the Gene Ontology annotations of the three main sub-ontologies (Biological Process, Molecular Function and Cellular Component) and from OMIM annotations [1], both represented as binary feature vectors, we constructed 4 more networks by using the classical Jaccard index to represent the edge weight (functional similarity) between the nodes (genes) of the resulting network. In our context the Jaccard index of two genes measures the ratio between the cardinality of their common annotations and the cardinality of the union of their annotations. The rationale behind the usage of this index is that two genes are similar if they share most of their annotations. All these annotations were obtained by parsing the raw text annotation files made available by Uniprot knowledge-base considering only its SWISSPROT component. Finally the resulting $n = 6$ networks have been integrated by averaging the edge weights w_{ij}^d between the genes i and j of each network $d \in \{1, n\}$ after normalizing their weights in the same range of values $w_{ij}^d \in [0, 1]$ (*Unweighted Average* (UA) network integration, [20]):

$$
\bar{w}_{ij} = \frac{1}{n} \sum_{d=1}^{n} w_{ij}^d \tag{7}
$$

The resulting weighted adjacency matrix representing the obtained networks is made up of 19,430 human proteins. From the HPO website we downloaded the January 2014 release, by considering the *Phenotypic Abnormality* subontology, that is the main subontology of the HPO (the other subontologies are significantly smaller and amount to only some tens of terms). To avoid prediction of HPO terms having too few annotations, for a reliable assessment we pruned HPO terms having less than 10 annotations obtaining a final HPO-*DAG* composed by 2154 HPO terms and 2641 between-terms-relationship.

The generalization performance of the methods were assessed through a classical 5-fold cross-validation procedure, whereas the results were evaluated by using the *gene-centric* metric F_{max} (i.e. the maximum hierarchical F-score achievable by "a posteriori" setting an optimal decision threshold [8]) and two *term-centric* metrics: the classical Area Under the Receiver Operating Characteristic Curve (*AUROC*) and the Area Under the Precision Recall Curve (*AUPRC*) to take into account the imbalance of annotated vs. unannotated HPO terms.

Table 1 summarizes the results achieved by the hierarchical methods HTD-DAG [19] and TPR-DAG [11] and by DESCENS, the novel ensemble variant presented in this manuscript.

Table 1. Average *AUROC* and *AUPRC* across terms and average F_{max}, Precision and Recall across genes of HTD-DAG, TPR-DAG and DESCENS ensemble variants using both *RANKS* and *SVMs* as base learner. Results of "flat" *RANKS* and *SVMs* are also reported. Results are estimated through 5-fold cross-validation. Separately for each metric and base learner the results significantly better than the others according to the Wilcoxon Rank Sum Test ($\alpha = 10^{-9}$) are highlighted in bold.

Method	AUROC	AUPRC	F_{max}	Precision	Recall
RANKS (flat)	0.8493	0.0910	0.3106	0.2407	0.4377
HTD-RANKS	0.8506	0.1065	0.3411	0.2717	0.4583
TPR-TF-RANKS	**0.8567**	0.1166	0.3547	0.2880	**0.4615**
TPR-T-RANKS	0.8512	0.1338	0.3574	0.2929	0.4582
TPR-W-RANKS	0.8507	0.1264	0.3620	0.3025	0.4506
DESCENS-TF-RANKS	0.8554	0.1082	0.3679	0.3148	0.4426
DESCENS-τ-RANKS	0.8530	**0.1360**	0.3622	0.3021	0.4520
DESCENS-T-RANKS	0.8503	0.1087	**0.3771**	**0.3227**	0.4535
DESCENS-W-RANKS	0.8502	0.1223	0.3671	0.3071	0.4561
SVM (flat)	0.7128	0.0429	0.1205	0.1165	0.1247
HTD-SVM	**0.8328**	0.0888	0.2597	0.1898	**0.4112**
TPR-TF-SVM	0.7060	0.0525	0.2034	0.1633	0.2694
TPR-T-SVM	0.8297	**0.1036**	0.2611	0.1939	0.3997
TPR-W-SVM	0.7915	0.0909	0.2187	0.1827	0.2723
DESCENS-TF-SVM	0.7092	0.0561	0.2338	0.1877	0.3100
DESCENS-τ-SVM	0.7182	0.0666	0.2424	0.1927	0.3266
DESCENS-T-SVM	0.7940	0.0514	**0.3102**	**0.2796**	0.3483
DESCENS-W-SVM	0.7724	0.0948	0.2373	0.1815	0.3427

In every experiment the hierarchical ensemble methods are able to improve the results of the flat methods used as base learner both in terms of $AUROC$, $AUPRC$ and F_{max}. More in detail looking at the results obtained using $RANKS$ as base learner, DESCENS-τ and DESCENS-T achieve better results than all the other compared methods in terms of $AUPRC$ and F_{max}, while TPR-TF achieves the best results in terms of $AUROC$, but HPO classes are highly imbalanced, and in this setting it is well-known that $AUPRC$ is a significantly more reliable metric than $AUROC$ [14]. Looking at the results obtained using as base learner the $SVMs$, we can observe that, independently of the ensemble method chosen, we achieve a significant strong improvement with respect to the flat prediction, especially in terms of $AUPRC$ and F_{max}. Interestingly enough, considering F_{max}, the only hierarchical metric among those considered, DESCENS achieves significantly better results both if we use $RANKS$ or $SVMs$ as base learners.

Finally we can observe that the performances of hierarchical ensembles largely depend on those of the flat base learners: for instance $DESCENS$-τ-$RANKS$ achieves a significantly higher precision at all recall levels with respect to $DESCENS$-W-SVM, due to the better performance of the $RANKS$ base learner (Fig. 3).

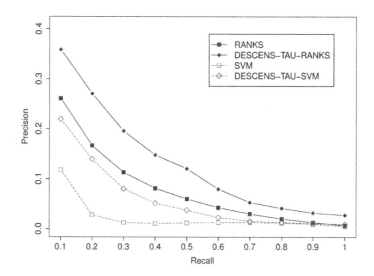

Fig. 3. Compared precision at different recall levels averaged across 2153 HPO terms of $DESCENS$-τ using $RANKS$ and SVM as base learner. The results of the corresponding flat methods, $RANKS$ and SVM are also reported.

This is not surprising since the improvement introduced by hierarchical ensemble methods also depends on the predictions of the underlying flat base learner: DESCENS can improve the flat predictions, but there is no guarantee of a correct prediction if most of the base flat learners provide incorrect predictions (Fig. 4).

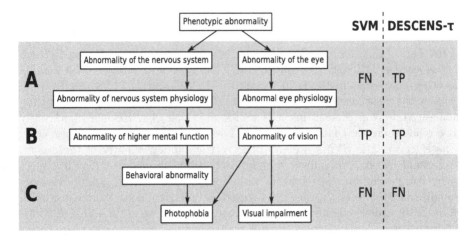

Fig. 4. Flat (*SVM*) and hierarchical *DESCENS-τ* HPO predictions for the gene *RGS9*. At the right side are displayed the correct *TP* and the incorrect *FN* predictions made respective by flat-*SVM* and by hierarchical *DESCENS-τ*. In the *A* box are depicted the predictions that the hierarchical method was able to correct with respect to flat method (*FN → TP*); in *B* are portrayed the correct predictions for both flat and hierarchical methods and finally in *C* are shown the incorrect flat predictions that the hierarchical method was not able to recover.

4   Conclusion

Genome and ontology wide experimental results show that the DESCENS algorithm is able to improve the predictions of both semi-supervised flat methods, such as the *RANKS* algorithm, that resulted one of the top ranked method in the recent *CAFA2* challenge for HPO term prediction [8], and of supervised methods such as *SVMs*, in terms of *AUROC*, *AUPRC* and F_{max}. Moreover DESCENS further improves HTD-DAG, and TPR-DAG, two of the state-of-the-art methods for HPO prediction, in terms of both *AUPRC* and F_{max}. Furthermore the proposed ensemble methods always provide consistent predictions that obey the *true path rule*, a fundamental fact to assure biologically coherent predictions among HPO terms.

Acknowledgments. We acknowledge partial support from the project "Discovering Patterns in Multi-Dimensional Data" (2016–2017) funded by Università degli Studi di Milano.

References

1. Amberger, J., Bocchini, C., Amosh, A.: A new face and new challenges for online mendelian inheritance in man (OMIM). Hum. Mutat. **32**, 564–7 (2011)
2. Ashburner, M., et al.: Creating the gene ontology resource: design and implementation. Genome Res. **11**(8), 1425–1433 (2001)

3. Bolstad, B.M., Irizarry, R.A., Astrand, M., Speed, T.P.: A comparison of normalization methods for high density oligonucleotide array data based on variance and bias. Bioinformatics **19**, 185–193 (2003)
4. Chatr-Aryamontri, A., et al.: The BioGRID interaction database: 2013 update. Nucleic Acids Res. **41**, 816–823 (2013)
5. Cormen, T., Leiserson, C., Rivest, R.L., Stein, S.: Introduction to Algorithms. MIT Press, Boston (2009)
6. Franceschini, A., et al.: STRING v9.1: protein-protein interaction networks, with increased coverage and integration. Nucleic Acids Res. **41**, 808–815 (2013)
7. Goldstein, B., Polley, E., Briggs, F.: Random forests for genetic association studies. Stat. Appl. Genet. Mol. Biol. **10**(1) (2011). https://doi.org/10.2202/1544-6115.1691
8. Jiang, Y., et al.: An expanded evaluation of protein function prediction methods shows an improvement in accuracy. Genome Biol. **17**, 184 (2016)
9. Kohler, S., Vasilevsky, N., Engelstad, M., et al.: The human phenotype ontology in 2017. Nucleic Acids Res. **45**, D865 (2017)
10. Moreau, Y., Tranchevent, L.: Computational tools for prioritizing candidate genes: boosting disease gene discovery. Nature Rev. Genet. **13**, 523–536 (2012)
11. Notaro, M., Schubach, M., Robinson, P.N., Valentini, G.: Prediction of human phenotype ontology terms by means of hierarchical ensemble methods. BMC Bioinform. **18**(1), 449:1–449:18 (2017). http://dblp.uni-trier.de/db/journals/bmcbi/bmcbi18.html#NotaroSRV17
12. Re, M., Mesiti, M., Valentini, G.: A fast ranking algorithm for predicting gene functions in biomolecular networks. IEEE/ACM Trans. Comput. Biol. Bioinf. **9**, 1812–1818 (2012)
13. Robinson, P.N., Frasca, M., Köhler, S., Notaro, M., Re, M., Valentini, G.: A hierarchical ensemble method for DAG-structured taxonomies. In: Schwenker, F., Roli, F., Kittler, J. (eds.) MCS 2015. LNCS, vol. 9132, pp. 15–26. Springer, Cham (2015). https://doi.org/10.1007/978-3-319-20248-8_2
14. Saito, T., Rehmsmeier, M.: The precision-recall plot is more informative than the ROC plot when evaluating binary classifiers on imbalanced datasets. PLOS ONE **10**, 1–21 (2015)
15. Schubach, M., Re, M., Robinson, P., Valentini, G.: Imbalance-aware machine learning for predicting rare and common disease-associated non-coding variants. Sci. Rep. **7**(2959) (2017). https://doi.org/10.1038/s41598-017-03011-5
16. Smedley, D., et al.: A whole-genome analysis framework for effective identification of pathogenic regulatory variants in Mendelian disease. Am. J. Hum. Genet. **99**, 595–606 (2016)
17. Valentini, G.: True Path Rule hierarchical ensembles for genome-wide gene function prediction. IEEE/ACM Trans. Comput. Biol. Bioinf. **8**, 832–847 (2011)
18. Valentini, G., Armano, G., Frasca, M., Lin, J., Mesiti, M., Re, M.: RANKS: a flexible tool for node label ranking and classification in biological networks. Bioinformatics **32**, 2872 (2016)
19. Valentini, G., Köhler, S., Re, M., Notaro, M., Robinson, P.N.: Prediction of human gene - phenotype associations by exploiting the hierarchical structure of the human phenotype ontology. In: Ortuño, F., Rojas, I. (eds.) IWBBIO 2015. LNCS, vol. 9043, pp. 66–77. Springer, Cham (2015). https://doi.org/10.1007/978-3-319-16483-0_7
20. Valentini, G., Paccanaro, A., Caniza, H., Romero, A., Re, M.: An extensive analysis of disease-gene associations using network integration and fast kernel-based gene prioritization methods. Artif. Intell. Med. **61**, 63–78 (2014)

21. Wang, P., et al.: Inference of gene-phenotype associations via protein-protein interaction and orthology. PLoS ONE **8**, 1–8 (2013)
22. Zemojtel, T., et al.: Effective diagnosis of genetic disease by computational phenotype analysis of the disease-associated genome. Sci. Transl. Med. **6**, 252ra123 (2014)

GP-Based Grammatical Inference for Classification of Amyloidogenic Sequences

Wojciech Wieczorek[1]([⊠]) [iD] and Olgierd Unold[2] [iD]

[1] Faculty of Computer Science and Materials Science,
University of Silesia in Katowice, Żytnia 12, 41-205 Sosnowiec, Poland
`wojciech.wieczorek@us.edu.pl`
[2] Department of Computer Engineering,
Wrocław University of Science and Technology,
Janiszewskiego 11/17, 50-372 Wrocław, Poland
`olgierd.unold@pwr.edu.pl`

Abstract. In this paper several methods of grammar induction problem are examined in the context of biological sequence analysis. In addition to this, a new method which generates noncircular context-free grammars is proposed. It has been shown through a computational experiment that the proposed, evolutionary-inspired approach overcomes statistically—with respect to classification quality—other grammatical inference algorithms on the sequences from a real amyloidogenic dataset.

Keywords: Grammatical inference ·
Noncircular context-free grammars · Genetic programming ·
Amyloidogenicity

1 Introduction

Protein sequences can be analyzed by means of diverse computational methods. One of their taxonomies groups the methods into five main categories: machine learning, alignment algorithms, statistical language learning, artificial intelligence (AI) techniques, and structural pattern recognition [5,7,8]. Unfortunately, there is no universal approach that can be successfully applied for all two-class bioinformatics datasets. Thus, every new idea may emerge as a valuable tool. For example, in hidden Markov models (HMMs) it is assumed that the probability of each symbol x_j depends only on a few previous symbols $x_{j-1}, \ldots, x_{j-m}$, which makes the model drastically simplified. Moreover, while parameters estimation in HMM is well refined, the choice of the model topology is a very hard task. In the case of support vector machines (SVMs), protein sequences have to be embedded into a d-dimensional feature space. A possible embedding is achieved by the relative frequencies of k-tuples of amino acids. This results in a feature

This research was supported by National Science Center, grant 2016/21/B/ST6/02158.

M. Bartoletti et al. (Eds.): CIBB 2017, LNBI 10834, pp. 81–91, 2019.
https://doi.org/10.1007/978-3-030-14160-8_9

space of dimension $d = 20^k$. Therefore, such an approach might also lead to too simple assumption or to the immense dimension of the input space. Additional examples can be multiplied, but it is clearly seen that every method has a certain fundamental weakness.

The purpose of the present proposal is twofold. Our main contribution is to provide a procedure for inferring the collection of noncircular context-free grammars (CFGs) which accepts all example sequences and a small number of counter-example sequences. The second objective is to determine that the proposed algorithm is better suited for the benchmark data—amyloidogenic sequence fragments—than selected comparative grammatical inference (GI) algorithms and a machine learning approach (SVM). The devised algorithm uses genetic programming, so it lies at the intersection of two categories: AI techniques and structural pattern recognition.

The most closely related works to our study are by Chirathamjaree and Ackroyd [3] and by Langdon and Barrett [9]. The former work is about generating compact noncircular context-free grammars having a near minimal number of rules and non-terminals, compatible with the requirement to be able to generate all sequences in the example set. Their work differs from ours in two respects. Firstly, they did not use the counter-example set. Secondly, in every rule of the form $A \rightarrow B\,C$ either B or C is the non-terminal corresponding to a letter (i.e. terminal symbol). The latter work is about using genetic programming (GP) to automatically create interpretable predictive models of a small number of very complex biological interactions of interest to medicinal and computational chemists who search for new drug treatments. Particularly, they found a simple predictive model of human oral bioavailability. Besides the difference in goals (we aim at obtaining a classifier for amyloidogenic sequences), another obvious distinction is the fact that our trees (basic GP structures) implicitly represent grammars and theirs represent arithmetic expressions. It is also worth mentioning our two recent works pertaining to classification of amyloidogenic sequences: one that uses directed acyclic word graphs [10] and one that uses star-free regular expressions [11]. These methods, however, cannot be applied to the present database since they are suitable for fixed-length sequences and they are too time-consuming for larger (thousands of sequences) datasets.

The paper's content is organized into five sections. In Sect. 2 we present necessary definitions and facts originating from combinatorics on words, formal languages and the GP method. Section 3 presents the proposed procedure of the construction of noncircular CFGs. Section 4 shows the experimental results of our approach against the background of the remaining GI algorithms. Conclusions and research perspectives are contained in Sect. 5.

2   Basic Definitions and Concepts

In this section, the concepts of string-rewriting systems, evolutionary computations and grammar induction problems are introduced. For more detailed definitions and advanced presentation of these topics, the reader is referred to [2, 8, 12] or other available textbooks.

2.1   Words, Grammars and Parse Trees

Grammars are very convenient tools to describe sets of sequences of symbols taken from a fixed alphabet. We will use the term *sequence* interchangeably with the term *word*. In the biological context, such words are usually sequences of the single-letter abbreviation of 20 amino acids. A *context-free grammar in Chomsky normal form* consists of a set of rules. Each rule is of one of the two possible forms: (1) $V_i \rightarrow V_j V_k$, or (2) $V_i \rightarrow A$, where Vs denote non-terminals and A denotes terminal (single-letter). Starting from a distinct non-terminal, let say V_0, we may generate words or check whether a given word belongs to a language (understood as a set of words) represented by the grammar. Such a production is performed via the series of rewritings. For example, the grammar: $V_0 \rightarrow V_1 V_2$, $V_1 \rightarrow V_1 V_1$, $V_1 \rightarrow A$, $V_2 \rightarrow V_2 V_2$, $V_2 \rightarrow C$ allows us to produce any word leading with As and ending with Cs. The following derivation: $V_0 \Rightarrow V_1 V_2 \Rightarrow A V_2 \Rightarrow A V_2 V_2 \Rightarrow A C V_2 \Rightarrow A C C$ proves that the word ACC can be produced (or in other words *is accepted*) by the grammar.

We say a CFG is *noncircular* if the right-hand side of every V_i's rule does not contain a non-terminal V_j, where $j \leq i$. The noncircularity of a grammar guarantees that the language accepted by the grammar is finite. From now on through the rest of this paper we will deal only with noncircular CFGs in Chomsky normal form, simply calling them grammars.

Every derivation can be presented as a *parse tree* whose internal nodes store non-terminals and leaves store terminals. The root of a tree is always V_0. If in a derivation there is $V_i \Rightarrow V_j V_k$ then in a corresponding tree piece there are two edges from V_i: bottom left to V_j and bottom right to V_k. If in a derivation there is $V_i \Rightarrow T$ then in a corresponding tree piece there is an edge from V_i down to T. A tree corresponding to the above-mentioned derivation of ACC is depicted in Fig. 1.

Fig. 1. A parse tree represented a derivation of the word ACC.

2.2   Genetic Programming

The genetic programming (GP) method used in our work is a kind of evolutionary computation, i.e., an algorithm for global optimization inspired by the natural microevolution of organisms. In technical terms, it is located in a family of population-based trial and error problem solvers with a stochastic optimization character. In evolutionary computation, an initial set of candidate solutions is generated and iteratively updated. Each new generation is produced by stochastically removing less desired solutions, and introducing small random changes. As a result, the population will gradually evolve to increase in fitness, in this case the chosen fitness function of the algorithm.

In GP the element of a population, so-called an individual, is usually a tree structure representing an expression. Random changes are performed through two operations: crossover (subtree exchange) and mutation (a single node modification). In our implementation, steady-state GP is used. This means that there are no generations. It differs from the generic evolutionary computation in such a way that instead of adding the children of the selected parents into the next generation, the two best individuals out of the two parents and two children are added back into the population so that the population size remains constant:

procedure run_GP:
    $P :=$ generate a population of p individuals randomly
    while stopping criterion has not been met:
        parent1, parent2 := tournament_selection(P)
        child1 := crossover(parent1, parent2)
        child2 := crossover(parent2, parent1)
        child1 := mutate(child1)
        child2 := mutate(child2)
        best1, best2 := get the two highest fitness
            individuals out of parent1, parent2, child1, child2
        replace parent1 with best1
        replace parent2 with best2

In tournament selection, k (the tournament size) individuals from the population are chosen at random and two of them with the highest fitness are returned. After obtaining a new individual in the initialization process or through the crossover or mutation operation, the individual is evaluated in order to determine its fitness f. The procedure uses the following conditions to determine when to stop: (i) steps (it stops when the number of loop iterations reaches an assumed value ℓ), (ii) time limit (it stops after running for an amount of time in seconds), (iii) fitness limit (it stops when the value of the fitness function for the best individual is lower than or equal to assumed fitness limit), or others. In experiments we used criterion (i).

2.3   Grammatical Inference

Let S_+ (examples) and S_- (counter-examples) be two (multi)sets of words over a finite alphabet Σ. From the viewpoint of the present research, grammatical inference (GI) problem relies on finding a compact grammar G that accepts the greatest proportion of S_+ and the tiniest proportion of S_-.

Note that more often it is required that G fulfills $S_+ \subseteq L(G)$ and $S_- \cap L(G) = \emptyset$, where $L(G)$ stands for a language accepted by G. A list of practical GI methods can be found in many works; the reader can refer to [12] as a good starting point on this topic.

3   The Present Method

Let Σ denote a finite alphabet. For given examples X and counter-examples Y which are words over Σ, our algorithm gives such a family of grammars, $\mathcal{G}$, that for every $x \in X$ there exists a $G \in \mathcal{G}$ satisfying $x \in L(G)$ and only the tiniest proportion of Y is accepted by any $G \in \mathcal{G}$.

procedure infer_grammars(X, Y):
    $\mathcal{G} := \emptyset$
    $U := X$ -- *uncovered words*
    while $U \neq \emptyset$:
        $S_+ :=$ choose with replacement p elements of U -- *a multiset*
        $S_- :=$ choose with replacement p elements of Y -- *a multiset*
        run_GP
        $t :=$ the fittest individual in P
        let $G(t)$ be a grammar extracted from a parse tree t
        if $L(G(t)) \cap U \neq \emptyset$:
            $U := U - L(G(t))$
            $\mathcal{G} := \mathcal{G} \cup G(t)$
    return $\mathcal{G}$ -- *the family of grammars*

In order to put the above procedure to work, we have to define the following elements and routines of GP: the primitives, the structure of an individual, the initialization, genetic operators, and the fitness function.

Individuals are parse trees composed of non-terminals, $V_0, V_1, \ldots, V_n$, and terminals being elements of Σ. An initial population is built upon S_+ so that each tree corresponds to one random derivation of a word. Naturally, trees must hold the following property: if V_j is below V_i in its left or right branch, then $j > i$. Crossover has been also designed in the way that satisfies this property. In Fig. 2 we can see two parent trees (the first two) and a child tree (the third), which is a combination of two parents made by substituting a part of one parent (shown in bold) with a part of the other (also shown in bold).

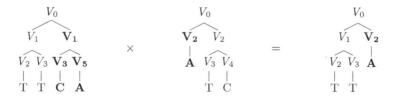

Fig. 2. An example of crossover.

More specifically, crossover proceeds as follows. Try these steps at most four times, otherwise return the copy of the first parent:

– Select a random internal node in each parent. Let say V_i and V_j.

– If $i \leq j$ then substitute the subtree selected by V_i in the copy of the first parent with the subtree selected by V_j in the copy of the second parent.

Mutation randomly decreases (if it is allowed with respect to noncircularity) the index of a non-terminal, in a randomly selected single node.

Finally, the fitness function measures a grammar's accuracy based on an individual t, and the sample (S_+, S_-) with Eq. 1:

$$f(t) = \frac{|\{w \in S_+ : w \in L(G(t))\}| + |\{w \in S_- : w \notin L(G(t))\}|}{2p}. \tag{1}$$

4   Referenced Methods and Results

We decided to compare our (GP) algorithm with four selected GI algorithms: two state merging algorithms, namely Blue-fringe and Traxbar, and two substring-based algorithms, ECGI (error-correcting grammatical inference), and ADIOS (automatic distillation of structure).

State merging algorithms are the most popular algorithms for the induction of finite state automata. Their common denominator is the merging of two states. All differences result from various particular reasons for choosing the pair of states to do this operation.

Substring-based algorithms is the group of algorithms for induction of automata and grammars in which words comparison for searching their fragments that are changed or fixed plays the crucial role. ECGI incrementally grows an inferred automaton by explicitly minimizing (with the help of dynamic programming) the number of states added when each new word is presented. ADIOS is a context-free grammar learning system, which relies on a statistical method for pattern extraction and on structured generalization. Please see [12] for a bibliographical references.

We also included one machine learning approach. An unsupervised data-driven distributed representation, called ProtVec [1], was applied and protein family classification has been performed using support vector machine classifier (SVM) [14] with the linear kernel.

In experiments we used the implementation[1] of our algorithm written in Python 3. An interpreter ran on an AMD Phenom II X6 1055T, 2.8 GHz processor under Windows 10 operating system with 8 GB RAM. The following values of parameters have been chosen: the size of a population, $p = 100$, the loop iteration limit, $\ell = 400$, the tournament size, $k = 7$, and the non-terminals limit, $n = 12$.

As a testbed, we considered the recently published AmyLoad dataset [13] containing a binary classification of almost 1500 unique, experimentally derived amyloidogenic sequence fragments. The analyses of such data are essential in studies of neurodegenerative diseases.

---

[1] https://github.com/wieczorekw/wieczorekw.github.io/tree/master/GP.

So as to estimate the GP's and compared approaches' ability to classify unseen sequences corrected repeated ten times tenfold cross-validation (10×10 cv) with 10 degrees of freedom test was applied. This scheme was proved to have excellent replicability [4] and, opposite to the MCNemar test and 5×2 cv test, low Type-II error. In repeated 10×10 cv, the data is randomly split in 10 mutually exclusive folds, next the model is learned over all but one fold and tested on the skipped fold. This protocol is repeated 10 times, each time assessing the model on the next skipped fold. To reduce variability, this scheme is repeated 10 times. Repeated r-times k-fold cross validation results in the following statistic

$$t_c = \frac{1/(k \cdot r) \sum_{i=1}^{k} \sum_{j=1}^{r} x_{ij}}{\sqrt{(1/(k \cdot r) + n_2/n_1)\hat{\sigma}^2}}$$

where x_{ij} is the difference of the performance quality between two compared algorithms on i-fold and j-run, n_1 is the number of instances used for training, n_2 the number of instances used for testing, $\hat{\sigma}^2$ is the variance of the n differences. To control the family-wise error rate, i.e. the probability that one or more Type-I errors will occur while multiple comparisons, the Holm correction was used [6].

To evaluate the quality classification of the compared models, we use the classification results stored in a confusion matrix. The following four scores were defined as tp, fp, fn, and tn, representing the numbers of true positives (correctly recognized amyloids), false positives (nonamyloids recognized as amyloids), false negatives (amyloids recognized as nonamyloids), and true negatives (correctly recognized nonamyloids), respectively. Based on the values stored in the confusion matrix, we calculate the widely used Precision and Recall, and combined metrics as F1-score, the AUC, and Matthews correlation coefficient.

Precision is defined as

$$P = tp/(tp + fp),$$

Recall as

$$R = tp/(tp + fn),$$

F1 as the harmonic mean of Precision and Recall

$$F1 = 2 \cdot (P \cdot R/(P + R)),$$

Area under Roc Curve (AUC) for binary classification as a Balanced Accuracy

$$AUC = (tp/p + tn/n)/2,$$

Matthews Correlation Coefficient is definded as

$$MCC = (tp \cdot tn - fp \cdot fn)/\sqrt{(tp + fn)(tp + fp)(tn + fp)(tn + fn)}.$$

Table 1. Performance of compared methods in terms of averaged Precision (P), Recall (R), F1, AUC, and MCC with standard deviation. The results are ordered by decreasing AUC.

	P	R	F1	AUC	MCC
GP	0.431 ± 0.064	0.561 ± 0.099	0.481 ± 0.057	0.619 ± 0.049	0.227 ± 0.094
Blue-fringe	0.467 ± 0.083	0.340 ± 0.065	0.391 ± 0.065	0.587 ± 0.039	0.193 ± 0.088
Traxbar	0.411 ± 0.061	0.413 ± 0.083	0.408 ± 0.059	0.580 ± 0.040	0.161 ± 0.081
SVM	0.571 ± 0.147	0.172 ± 0.064	0.260 ± 0.086	0.559 ± 0.034	0.187 ± 0.101
ECGI	0.787 ± 0.182	0.119 ± 0.050	0.205 ± 0.079	0.554 ± 0.027	0.230 ± 0.095
ADIOS	0.328 ± 0.071	0.628 ± 0.228	0.402 ± 0.083	0.524 ± 0.051	0.048 ± 0.109

Comparative analysis of the five measures (Precision, Recall, F-score, the AUC, and Matthews correlation coefficient) is summarized in Table 1 and Figs. 3, 4 and 5. These quantities are reported for six compared predictors. Reported numerical results show that GP has the highest values in all but one combined measures (i.e. AUC and F1) of the performance of a binary classification test. Table 2 gives adjusted by Holm procedure p values for the comparison of the GP as the control method with the remaining algorithms. These p values indicate that there are significant performance differences between GP algorithm and all compared methods over AUC and F1 measures. Additionally, GP outperforms statistically Traxbar and ADIOS approaches regarding MCC measure.

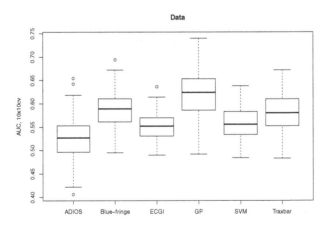

Fig. 3. Performance comparison of ADIOS, Blue-fringe, ECGI, GP, Traxbar, and SVM methods. Boxplots represent the AUC values obtained from 10 × 10 cross-validation.

Table 2. p-values for the comparison of the GP as a control method with the other methods and F1, AUC, and MCC measures. The initial level of confidence $\alpha = 0.05$ is adjusted by Holm procedure.

GP versus	Holm p for F1	Holm p for AUC	Holm p for MCC
Blue-fringe	9.036106 e−11	5.219685 e−04	1.505537 e−01
Traxbar	1.833783 e−08	6.211608 e−05	1.068778 e−03
ECGI	3.508577 e−35	4.400321 e−12	8.360555 e−01
ADIOS	2.238259 e−07	1.210657 e−14	5.612043 e−13
SVM	9.377921 e−26	1.409218 e−09	1.474196 e−01

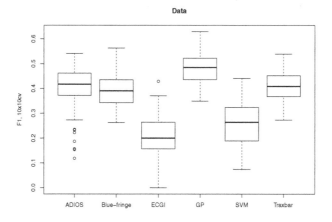

Fig. 4. Performance comparison of ADIOS, Blue-fringe, ECGI, GP, Traxbar, and SVM methods. Boxplots represent the F1 values obtained from 10×10 cross-validation.

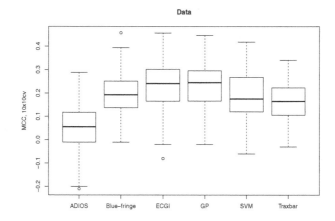

Fig. 5. Performance comparison of ADIOS, Blue-fringe, ECGI, GP, and Traxbar methods. Boxplots represent the MCC values obtained from 10×10 cross-validation.

5   Conclusions and Future Work

This paper dealt with the induction of noncircular CFGs based on a finite sample. That is constituted by the following task: given two sets X, Y of words build a family of compact grammars which is well fitting with the input (recognizes X and rejects the most of Y). In order to address the problem, we have made use of genetic programming. The experiments conducted showed that the new proposed algorithm outperformed statistically all the compared methods in terms of F-score and AUC for real amyloidogenic sequences.

Our method has also a few disadvantages. The most significant ones are: sensitiveness to parameters' values and long working time. In the near future, we are planning to minimize the execution time by considering parallel implementation.

References

1. Asgari, E., Mofrad, M.R.K.: Continuous distributed representation of biological sequences for deep proteomics and genomics. PLoS ONE **10**(11), e0141287 (2015). https://doi.org/10.1371/journal.pone.0141287
2. Banzhaf, W., Francone, F.D., Keller, R.E., Nordin, P.: Genetic Programming: An Introduction: On the Automatic Evolution of Computer Programs and Its Applications. Morgan Kaufmann, San Francisco (1998)
3. Chirathamjaree, C., Ackroyd, M.H.: A method for the inference of non-recursive context-free grammars. Int. J. Man Mach. Stud. **12**(4), 379–387 (1980)
4. Bouckaert, R.R., Frank, E.: Evaluating the replicability of significance tests for comparing learning algorithms. In: Dai, H., Srikant, R., Zhang, C. (eds.) PAKDD 2004. LNCS, vol. 3056, pp. 3–12. Springer, Heidelberg (2004). https://doi.org/10.1007/978-3-540-24775-3_3
5. Durbin, R., Eddy, S., Krogh, A., Mitchison, G.: Biological Sequence Analysis: Probabilistic Models of Proteins and Nucleic Acids. Cambridge University Press, New York (1998)
6. Holm, S.: A simple sequentially rejective multiple test procedure. Scand. J. Stat. **6**, 65–70 (1979)
7. Hu, X., Pan, Y.: Knowledge Discovery in Bioinformatics: Techniques, Methods, and Applications. Wiley, New Jersey (2007)
8. Keedwell, E., Narayanan, A.: Intelligent Bioinformatics: The Application of Artificial Intelligence Techniques to Bioinformatics Problems. Wiley, Chichester (2005)
9. Langdon, W.B., Barrett, S.J.: Genetic programming in data mining for drug discovery. In: Ghosh, A., Jain, L.C. (eds.) Evolutionary Computing in Data Mining, vol. 163, pp. 211–235. Springer, Heidelberg (2005). https://doi.org/10.1007/3-540-32358-9_10
10. Wieczorek, W., Unold, O.: Induction of directed acyclic word graph in a bioinformatics task. In: JMLR Workshop and Conference Proceedings, vol. 34, pp. 207–217 (2014)
11. Wieczorek, W., Unold, O.: Use of a novel grammatical inference approach in classification of amyloidogenic hexapeptides. Comput. Math. Methods Med. **2016** (2016). Article ID 1782732

12. Wieczorek, W.: Grammatical Inference: Algorithms, Routines and Applications. Springer, Cham (2017). https://doi.org/10.1007/978-3-319-46801-3
13. Wozniak, P.P., Kotulska, M.: AmyLoad: website dedicated to amyloidogenic protein fragments. Bioinformatics **31**(20), 3395–3397 (2015)
14. Wu, T.-F., Lin, C.-J., Weng, R.C.: Probability estimates for multi-class classification by pairwise coupling. J. Mach. Learn. Res. **5**, 975–1005 (2004)

Estimation of Kinetic Reaction Constants: Exploiting Reboot Strategies to Improve PSO's Performance

Simone Spolaor[1(✉)] [ID], Andrea Tangherloni[1] [ID], Leonardo Rundo[1,2] [ID],
Paolo Cazzaniga[3,4] [ID], and Marco S. Nobile[1,3] [ID]

[1] Department of Informatics, Systems and Communication,
University of Milano-Bicocca, 20126 Milano, Italy
{simone.spolaor,andrea.tangherloni,leonardo.rundo,nobile}@disco.unimib.it
[2] Institute of Molecular Bioimaging and Physiology,
Italian National Research Council, 90015 Cefalú, PA, Italy
[3] SYSBIO.IT Centre of Systems Biology, 20126 Milano, Italy
[4] Department of Human and Social Sciences, University of Bergamo,
24129 Bergamo, Italy
paolo.cazzaniga@unibg.it

Abstract. The simulation and analysis of mathematical models of biological systems require a complete knowledge of the reaction kinetic constants. Unfortunately, these values are often difficult to measure, but they can be inferred from experimental data in a process known as Parameter Estimation (PE). In this work, we tackle the PE problem using Particle Swarm Optimization (PSO) coupled with three different reboot strategies, which aim to reinitialize particle positions to avoid local optima. In particular, we highlight the better performance of PSO coupled with the reboot strategies with respect to standard PSO. Finally, since the PE requires a huge number of simulations at each iteration of PSO, we exploit cupSODA, a GPU-powered deterministic simulator, which performs all simulations and fitness evaluations in parallel.

Keywords: Particle Swarm Optimization · Parameter Estimation · GPGPU computing · cupSODA · Systems Biology

1 Scientific Background

Mechanism-based mathematical models are used in the field of Systems Biology to provide a detailed description of the biochemical processes occurring in living cells [21]. A complete parameterization of such models is needed to perform simulations of the dynamics; however, kinetic parameters are often difficult to measure through experimental methodologies. Different *Parameter Estimation* (PE) techniques [11], relying either on approximation strategies or global optimization methods, can be employed to automatically infer a parameterization that fits the experimental data. Especially, *Particle Swarm Optimization* (PSO)

© Springer Nature Switzerland AG 2019
M. Bartoletti et al. (Eds.): CIBB 2017, LNBI 10834, pp. 92–102, 2019.
https://doi.org/10.1007/978-3-030-14160-8_10

has shown to be one of the best optimization techniques for the PE [2,5,7,12]. PSO is a meta-heuristic in which a population (*swarm*) of candidate solutions (*particles*) explores a bounded search space to find the best solution to a certain problem, according to a fitness function.

Different works presented variants aimed to improve the standard PSO algorithm, by introducing specific strategies that attempted to overcome stagnation of the swarm in local minima, either by implementing hybrid PSO approaches [18,23] or reboot strategies [6,8]. In this work, we consider different reboot strategies to re-initialize particles if they are too close to the global best position or if the global or local best positions remain unchanged for a given number of iterations, and assess their influence on the performance of PSO applied to the PE of the kinetic constants of biochemical systems. Our results show that rebooting can considerably improve PSO's performances, in particular when a "local" methodology (i.e., based on local best position monitoring) is exploited.

In this context, the PE typically requires a huge amount of simulations during each iteration of PSO for the fitness calculation; we therefore exploit a GPU-powered deterministic simulator, called cupSODA, to efficiently perform in parallel all simulations, strongly reducing the required computation time.

The paper is structured as follows. In Sect. 2 we introduce the parameter estimation problem and the basic notions of PSO, we describe the reboot strategies exploited in this study, as well as briefly recall the features of cupSODA and of the parallelization strategy adopted in this work. Section 3 presents the results of performance analysis regarding the PE on the mathematical model of the Heat Shock Response (HSR) mechanism. Finally, in Sect. 4 we conclude with some final remarks and directions for future work.

2 Materials and Methods

2.1 Parameter Estimation of Biochemical Systems

In this study, we focus on mechanistic models of biochemical systems, defined according to the reaction-based formalism. Reaction-based models consist of a set $S = \{S_1 \ldots, S_N\}$ containing the molecular species occurring in the system, and a set $\mathcal{R} = \{R_1, \ldots, R_M\}$ of reactions describing the interactions that take place among them. A reaction R_j, with $j = 1, \ldots, M$, is characterized by the sets of its reactants and products, whose elements belong to S, and a kinetic constant $k_j \in \mathbb{R}^+$, whose unit of measure depends on the reaction order (i.e., the number of reacting molecules). Since kinetic constants' values are typically hard or impossible to measure experimentally, hereafter we suppose to have zero or partial knowledge about the vector of kinetic constants $\mathbf{k} = (k_1, \ldots, k_M)$, while we assume to have complete knowledge of the sets S and $\mathcal{R}$, as well as of the initial concentrations of the molecular species. Note that, in principle, the PE strategy presented in this paper can be applied to estimate the initial concentrations of the molecular species involved in the model.

A PE can be performed if some target data are available, such as the measurement of the molecular amount of a subset of species $S' = \{S_1, \ldots, S_S\} \subseteq S$, with

$S \leq N$. In this context, each element of $\mathcal{S}'$ corresponds to a time-series, which consists of experimental data sampled at a finite set of time points $\tau_1, \ldots, \tau_F$. $Y_s(\tau_f)$ denotes the molecular amount of species $S_s \in \mathcal{S}'$ measured at time τ_f ($s = 1, \ldots, S$ and $f = 1, \ldots, F$). Overall, this set of experimental measurements is called *discrete-time target series* (DTTS).

The PE process consists in comparing the DTTS of every $S_s \in \mathcal{S}'$ with the simulated dynamics of the same species S_s, obtained by executing an *in silico* simulation of the model. In particular, in this work we adopted cupSODA [13], a deterministic simulator that takes as input the model, the initial amounts of all species in $\mathcal{S}$, and the candidate solutions $\boldsymbol{k} = (k_1, \ldots, k_D)$, with $D \leq M$, containing the kinetic constants of the reactions in $\mathcal{R}' = \{R_1, \ldots, R_D\} \subseteq \mathcal{R}$, which need to be estimated. Given the outcome of a cupSODA simulation, we sample a set of consecutive time instants $\tau_0, \ldots, \tau_{max}$, where τ_0 and τ_{max} are the first and last instant of the simulation, respectively. We denote by $X_s^k(\tau)$ the molecular amount of the species $S_s \in \mathcal{S}'$ at time τ, with $s = 1, \ldots, S$ and $\tau_0 \leq \tau \leq \tau_{max}$.

The fitness of a candidate solution vector $\boldsymbol{k}$ is computed by comparing the measured experimental data of species S_s in the DTTS with the simulation outcome: (i) we determine the value of $X_s^k(\tau_f)$, that is, the simulated molecular amount of S_s at time point τ_f, for each $f = 1, \ldots, F$; (ii) we calculate how much the simulated molecular amounts $X_s^k(\tau_f)$ match the experimental measures $Y_s(\tau_f)$ at each time instant $f = 1, \ldots, F$, for each species $S_s \in \mathcal{S}'$. To be more precise, the fitness function is defined as follows:

$$\mathcal{F}(\boldsymbol{k}) = \sum_{f=1}^{F} \sum_{s=1}^{S} \frac{|Y_s(\tau_f) - X_s^k(\tau_f)|}{Y_s(\tau_f)}. \tag{1}$$

$\mathcal{F}(\boldsymbol{k})$ measures the distance between the DTTS and the simulation outcome of all species in $\mathcal{S}'$ and all sampled time instants τ_f. Therefore, the PE problem consists in minimizing $\mathcal{F}(\boldsymbol{k})$ to identify the vector $\boldsymbol{k}$ able to provide a simulated dynamics that matches at best the DTTS.

2.2   Particle Swarm Optimization

In PSO, a set of candidate solutions is represented by a population of n particles (*swarm*). Each particle i, where $i = 1, \ldots, n$, is characterized by a position vector $\mathbf{x}_i$ and a velocity vector $\mathbf{v}_i$ that allow to update its own current position, both defined in a multi-dimensional search space $\mathbb{R}^D$. During the optimization process, particles are attracted towards the best position $\mathbf{b}_i \in \mathbb{R}^D$, found by the particle itself, and the global best position $\mathbf{g} \in \mathbb{R}^D$, reached by the entire swarm, until a termination criterion is achieved.

Global exploration and local exploitation behaviors are governed by two parameters: the social factor $c_{soc} \in \mathbb{R}^+$ and the cognitive factor $c_{cog} \in \mathbb{R}^+$, respectively. Furthermore, two vectors $\mathbf{r}_1, \mathbf{r}_2$ of random numbers, sampled from a uniform distribution in $[0, 1)$, are employed to add stochasticity to particles movement and prevent premature convergence to local optima. A maximum velocity vector $\mathbf{v}_{max} = (v_{max_1}, \ldots, v_{max_D})$, with $v_{max_d} \in \mathbb{R}^+$ and $d = 1, \ldots, D$, limits the velocity of particles that are also modulated by an inertia factor $w \in \mathbb{R}^+$ to avoid chaotic behaviors in the swarm. In this work, we assess the effects of fixed and linearly decreasing inertia on the performance of PSO in solving a PE problem. For the inertia parameter, we set the values typically used in literature (such as in [2]) in which the inertia is either kept fixed to $w = 0.729$, or decreases from 0.9 down to 0.4. According to this formulation, the velocity of each particle i is updated as $\mathbf{v}_i = w \cdot \mathbf{v}_i + c_{soc} \cdot \mathbf{r}_1 \circ (\mathbf{x}_i - \mathbf{g}) + c_{cog} \cdot \mathbf{r}_2 \circ (\mathbf{x}_i - \mathbf{b}_i)$, where $\circ$ denotes the component-wise product. Lastly, the particle position $\mathbf{x}_i$ is updated as $\mathbf{x}_i = \mathbf{x}_i + \mathbf{v}_i$.

In PSO, the search space is bounded to avoid the divergence of candidate solutions towards infinity. We indicate as β_d^{min} and β_d^{max} the lower and upper bound of the d-th dimension, respectively, with $d = 1, \ldots, D$. Moreover, in order to keep the particles inside the bounded region, we assume damping boundary conditions, in which the particle "bounces" (with a random elasticity) on the limit of the search space. Particle initialization is performed by sampling from a logarithmic distribution, as described in [3].

2.3 Reboot Strategies in PSO

Reboot strategies for particles have been introduced to avoid early convergence of the swarm and stagnation in local minima. A reboot strategy should accurately and efficiently identify particles that are no longer exploring effectively the search space, and re-initialize them to increase diversity in the swarm. In the context of continuous optimization problems in large search spaces, different reboot strategies were integrated in PSO [6,8], which consist in "restarting" the algorithm by re-initializing all particles in the swarm except for the global best position found so far.

In this work, we exploit reboot strategies and assess their influence on the performance of PSO applied to the PE of the kinetic constants of biochemical systems. The following re-initialization approaches were implemented:

- (i) *Global*, if the global best position $\mathbf{g}$ of the swarm does not improve for η iterations (where η is a user-defined parameter), each particle of the swarm is randomly re-initialized in the logarithmic interval $(\beta_d^{min}, \beta_d^{max})$, its local best position $\mathbf{b}_i$ is set to the new position $\mathbf{x}_i$ and its velocity $\mathbf{v}_i$ is reset to $\mathbf{0}$;
- (ii) *Local*, if the local best position $\mathbf{b}_i$ of particle i does not improve for η iterations, its position $\mathbf{x}_i$ is randomly re-initialized in the logarithmic interval $(\beta_d^{min}, \beta_d^{max})$, its local best position $\mathbf{b}_i$ is set to the new random position and its velocity $\mathbf{v}_i$ is reset to $\mathbf{0}$;
- (iii) *Distance*, if $\|\mathbf{b}_i^* - \mathbf{g}^*\| < \theta$, particle i is randomly re-initialized in the logarithmic interval $(\beta_d^{min}, \beta_d^{max})$, its local best position $\mathbf{b}_i$ is set to $\mathbf{x}_i$ and its velocity $\mathbf{v}_i$ is reset to $\mathbf{0}$. θ is a user setting, $\mathbf{b}_i^* = (\log_{10}(b_{i,1}), \ldots, \log_{10}(b_{i,D}))$ and $\mathbf{g}^* = (\log_{10}(g_1), \ldots, \log_{10}(g_D))$ are the logarithmic transformation of $\mathbf{b}_i$ and $\mathbf{g}$, respectively.

2.4   Deterministic Simulations on GPUs Using cupSODA

cupSODA [13] is a GPU-based simulator capable of accelerating the tasks typically executed in the field of Systems Biology—such as PE, sensitivity analysis or reverse engineering [1,4]—which are usually computationally intensive since they involve large batches of simulations of a model. cupSODA exploits an efficient coarse-grained parallelization strategy to perform deterministic simulations, by launching a GPU thread for each model parameterization, running in parallel a huge amount of independent simulations of the same model. In particular, each simulation is characterized by a different setting of initial molecular amounts of chemical species and/or kinetic constants values.

cupSODA is built upon a C version of the numerical integrator LSODA [20], ported and adapted to the CUDA architecture [17]. cupSODA is a *black-box* simulator that can be easily employed without any programming skills by the final user: given a reaction-based model of a biochemical system, cupSODA automatically generates the corresponding system of ODEs, according to the mass-action kinetics [24]. In order to reduce the memory latencies [16], cupSODA exploits the shared memory of the GPU to save the current state of each simulation, and the constant memory of the GPU to save all the constants values (e.g., number of reactions and chemical species in the model, length of ODEs and Jacobian arrays, etc.) and the LSODA settings [9].

Considering the specific case of the PE, cupSODA can offer an additional feature, as it allows for easily comparing the outcome of simulations with any available experimental data. In particular, given a set of F time points sampled in a DTTS, cupSODA proceeds as follows: (i) it invokes the LSODA kernel $F - 1$ times; (ii) at each time, the kernel is run over a time window of length $\Delta\tau = \tau_f - \tau_{f-1}$, $f = 1, \ldots, F$; (iii) at the end of each $\Delta\tau$, it stores the simulated molecular amounts of each species in $\mathcal{S}'$. Finally, cupSODA calculates in parallel the fitness values associated with candidate model parameterizations, following Eq. (1).

3   Results

In this section, we compare the performances of the reboot strategies described in Sect. 2.3 with respect to the standard PSO. In all tests, we used the following settings for PSO: (i) damping boundary conditions; (ii) maximum velocity $v_{max_d} = 0.2 \cdot \psi_d$, where $\psi_d = \beta_d^{max} - \beta_d^{min}$, for $d \in \{1, 2, \ldots, D\}$; ($iii$) number of iterations $\text{IT}_{\text{MAX}} = 1000$. We also set $c_{cog} = c_{soc} = 1.494$, which correspond to the values typically used in literature, and compared the effect of decreasing inertia (from $w = 0.9$ to $w = 0.4$) against a fixed value ($w = 0.729$) (see [22] for additional details).

The performances are evaluated by computing the *average best fitness* (ABF) over $\Phi = 20$ independent repetitions of the PE, and by showing the box-plots obtained with the best fitness values reached by PSO at the last iteration of each PE run. Note that, since we are minimizing the fitness value, a smaller ABF corresponds to a better result.

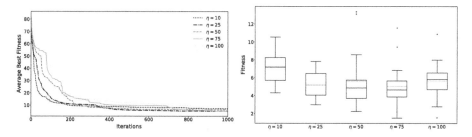

Fig. 1. *Left*: ABF achieved by PSO implementing the *Global* reboot strategy with different η values. In all cases, we used $w = 0.729$ and $n = 512$. *Right*: box-plots obtained considering the best fitness values reached by PSO with *Global* reboot at the last iteration of each optimization, with the same settings specified above. The solid (dashed) line corresponds to the median (mean).

All tests were executed exploiting a model of the Heat Shock Response (HSR) presented in [19], consisting of 17 reactions among 10 species. HSR is a regulatory mechanism that allows the cell to quickly react to high temperatures and other forms of physiological and environmental stress. The initial molecular amounts of the species are also given in [19]. In the PE of the HSR model, we consider $D = 16$ dimensions, since the value of a kinetic constant is known and kept to its real value throughout the optimization. The search space boundaries were fixed to $\beta_d^{min} = 10^{-10}$ and $\beta_d^{max} = 100$, for each $d = 1, \ldots, D$. The target data used in the PE were generated *in silico*, by sampling 140 points from a simulation realized by using a reference kinetic parameterization.

First of all, we considered the influence of the swarm size and of the inertia value on the standard PSO algorithm. In particular, we executed the PE using: a swarm composed of $n = 32$ particles and fixed inertia $w = 0.729$; a swarm with $n = 512$ particles and a linearly decreasing inertia (from $w = 0.9$ to $w = 0.4$ throughout the entire simulation); a swarm with $n = 512$ particles and inertia $w = 0.729$. These results (data not shown) highlighted that: (i) as expected, a large number of particles allows us to achieve lower ABF values and reach better results at the end of the optimization process with respect to a swarm of smaller size; (ii) a fixed inertia value outperforms PSO with a decreasing inertia in this particular test case.

Figures 1, 2 and 3 illustrate the results obtained by PSO implementing the three different reboot strategies. In these tests, we varied the threshold parameters exploited by the reboot strategies. Especially, we considered the values $10, 25, 50, 75, 100$ of the parameter η of *Global* and *Local* reboot strategies, and the values $0.01, 0.1, 0.25, 0.5, 0.75, 1.0$ of the parameter θ of the *Distance* reboot strategy. Note that these values were arbitrarily chosen to the aim of assessing the influence of η and θ on the performance of PSO coupled with reboot strategies.

On the one hand, the performance of the *Distance* reboot (Fig. 3) appears to be affected by the choice of θ values; on the other hand, both *Global* reboot

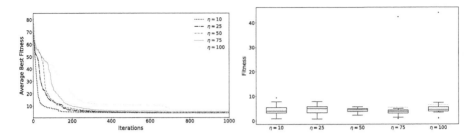

Fig. 2. *Left*: ABF achieved by PSO implementing the *Local* reboot strategy with different η values. In all cases, we used $w = 0.729$ and $n = 512$. *Right*: box-plots obtained considering the best fitness values reached by PSO with *Local* reboot at the last iteration of each optimization, with the same settings specified above. The solid (dashed) line corresponds to the median (mean).

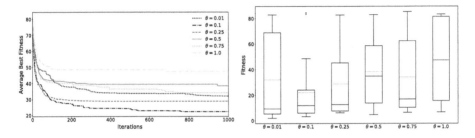

Fig. 3. *Left*: ABF achieved by PSO implementing the *Distance* reboot strategy with different θ values. In all cases, we used $w = 0.729$ and $n = 512$. *Right*: box-plots obtained considering the best fitness values reached by PSO with *Distance* reboot at the last iteration of each optimization, with the same settings specified above. The solid (dashed) line corresponds to the median (mean).

(Fig. 1) and *Local* (Fig. 2) strategies are robust with respect to the choice of η values, a result confirmed by the smaller dispersion in the box-plots. It is worth noting that the threshold parameters directly affect the number of reboots occurring at each iteration of PSO, even though this does not necessarily lead to better performances. For instance, it is evident how $\theta = 0.1$ performs better than both $\theta = 0.01$ and $\theta = 0.25$ in the case of *Distance* reboot (Fig. 3). Therefore, a good trade-off for the threshold parameters should be identified to prevent excessive or insufficient rebooting of particles.

Afterwards, we selected the best threshold values of each reboot strategy and compared them against standard PSO. Figure 4 shows that all reboot strategies outperform standard PSO both in terms of ABF and interquartile range of the final best fitness; this experimental evidence is observed especially in the case of *Global* and *Local* reboot strategies.

Since the *Local* strategy seemed to consistently outperform the other approaches, we tested an additional reboot strategy: instead of sampling the new positions from a logarithmic distribution, we used a lognormal distribution

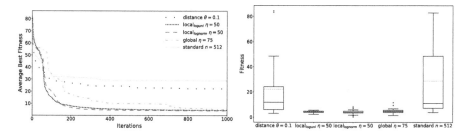

Fig. 4. *Left*: ABF achieved by the standard PSO ($n = 512$ and $w = 0.729$) and PSO with the three reboot strategies, performed with their respective best threshold values. *Right*: box-plots obtained considering the best fitness values reached at the last iteration of each optimization ϕ, with the same settings specified above. The solid (dashed) line corresponds to the median (mean).

with mean equal to $(\log(\beta_d^{min}) + \log(\beta_d^{max}))/2$ and σ equal to $(\log(\beta_d^{max}) - \log(\beta_d^{min}))/2$ [10]. In Fig. 4 we denote the *Local* strategy based on logarithmic samples and lognormal samples with $\texttt{local}_{loguni}$ and $\texttt{local}_{lognorm}$, respectively. According to our results, the new version slightly improves the average performances (left plot) although with a higher variance in the final ABF (right plot). Moreover, the generation of lognormally distributed random deviates is more computationally expensive than uniform random numbers (approximately five times slower). Due to these drawbacks, $\texttt{local}_{loguni}$ remains the most reliable strategy for PSO reboots.

Finally, in order to verify the correctness of the PE, we simulated the dynamics of the HSR model by using the best parameterization found by PSO (over the 20 runs) with *Local* and *Global* reboot strategies. Figure 5 shows that both the *Local* and *Global* reboot allowed to simulate a dynamics that closely approximates the target temporal dynamics of the hsf$_3$:hse species.

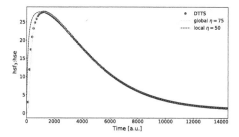

Fig. 5. Temporal dynamics of the hsf$_3$:hse species of the HSR model, obtained with the best parameterization found by PSO coupled with the *Local* ($\eta = 50$) and *Global* ($\eta = 75$) reboot strategies.

4   Conclusion

In this paper, we considered different reboot strategies for particles in PSO and assessed their performance for the estimation of kinetic constants in reaction-based models of biological systems. PE might be computationally intensive even in the case of small-scale models of biological systems, as it requires a huge amount of fitness function evaluations. We therefore exploited cupSODA [13], a GPU-powered deterministic simulator that performs large batches of simulations relying on a coarse-grained parallelization strategy, which allowed to speedup the computation up to 86× with respect to a CPU-based execution.

The results obtained in this work highlight that reboot strategies can be successfully coupled with PSO to avoid early convergence and stagnation of the swarm. In particular, the *Global* and *Local* reboot strategies proved to be considerably reliable, since they are characterized by (i) high convergence speed, (ii) lower ABF with respect to the *Distance* reboot strategy, and (iii) higher robustness with respect to different choices for their functioning setting η.

As a future extension of this work, we plan to apply the PE methodology based on PSO and cupSODA to more complex models of biological systems, where standard approaches fail because of the computational burden of the simulation process. We will also consider a multi-swarm approach [14], where a set of DTTS—measured in different experimental conditions—are simultaneously taken into account to the aim of improving the quality and the biological relevance of the inferred model parameterization. In this situation, special reboot methods can be investigated to implement asynchronous migration strategies between the swarms.

Finally, when the biochemical systems under investigation are characterized by molecular species present in small quantities, stochasticity plays a fundamental role and cannot be neglected. In such a situation, stochastic simulation algorithms must be employed to calculate the fitness of PSO particles; we will therefore replace cupSODA with cuTauLeaping [15], a GPU-based stochastic simulator, and we will develop an appropriate fitness function capable of dealing with complex and noisy dynamics.

Acknowledgement. We gratefully acknowledge the support of NVIDIA Corporation with the donation of the GeForce GTX Titan X GPU used for this research.

This work was conducted in part using the resources of the Advanced Computing Center for Research and Education at Vanderbilt University, Nashville, TN, USA.

References

1. Aldridge, B.B., Burke, J.M., Lauffenburger, D.A., Sorger, P.K.: Physicochemical modelling of cell signalling pathways. Nat. Cell Biol. **8**, 1195–1203 (2006)
2. Besozzi, D., Cazzaniga, P., Mauri, G., Pescini, D., Vanneschi, L.: A comparison of genetic algorithms and particle swarm optimization for parameter estimation in stochastic biochemical systems. In: Pizzuti, C., Ritchie, M.D., Giacobini, M. (eds.) EvoBIO 2009. LNCS, vol. 5483, pp. 116–127. Springer, Heidelberg (2009). https://doi.org/10.1007/978-3-642-01184-9_11

3. Cazzaniga, P., Nobile, M.S., Besozzi, D.: The impact of particles initialization in PSO: parameter estimation as a case in point. In: Proceedings of IEEE Conference on Computational Intelligence in Bioinformatics and Computational Biology (CIBCB), pp. 1–8 (2015)
4. Chou, I.C., Voit, E.O.: Recent developments in parameter estimation and structure identification of biochemical and genomic systems. Math. Biosci. **219**(2), 57–83 (2009)
5. Da Ros, S., et al.: A comparison among stochastic optimization algorithms for parameter estimation of biochemical kinetic models. Appl. Soft Comput. **13**(5), 2205–2214 (2013)
6. De Oca, M.A.M., Stutzle, T., Birattari, M., Dorigo, M.: Frankenstein's PSO: a composite particle swarm optimization algorithm. IEEE Trans. Evol. Comput. **13**(5), 1120–1132 (2009)
7. Dräger, A., Kronfeld, M., Ziller, M.J., Supper, J., Planatscher, H., Magnus, J.B.: Modeling metabolic networks in *C. glutamicum*: a comparison of rate laws in combination with various parameter optimization strategies. BMC Syst. Biol. **3**(5) (2009)
8. García-Nieto, J., Alba, E.: Restart particle swarm optimization with velocity modulation: a scalability test. Soft Comput. **15**(11), 2221–2232 (2011)
9. Harris, L.A., et al.: GPU-powered model analysis with PySB/cupSODA. Bioinformatics **33**(21), 3492–3494 (2017). (btx420)
10. Limpert, E., Stahel, W.A., Abbt, M.: Log-normal distributions across the sciences: keys and clues. BioScience **51**(5), 341–352 (2001)
11. Mendes, P., Kell, D.: Non-linear optimization of biochemical pathways: applications to metabolic engineering and parameter estimation. Bioinformatics (Oxford, England) **14**(10), 869–883 (1998)
12. Moles, C.G., Mendes, P., Banga, J.R.: Parameter estimation in biochemical pathways: a comparison of global optimization methods. Genome Res. **13**(11), 2467–2474 (2003)
13. Nobile, M.S., Besozzi, D., Cazzaniga, P., Mauri, G.: GPU-accelerated simulations of mass-action kinetics models with cupSODA. J. Supercomput. **69**(1), 17–24 (2014)
14. Nobile, M.S., Besozzi, D., Cazzaniga, P., Mauri, G., Pescini, D.: A GPU-based multi-swarm PSO method for parameter estimation in stochastic biological systems exploiting discrete-time target series. In: Giacobini, M., Vanneschi, L., Bush, W.S. (eds.) EvoBIO 2012. LNCS, vol. 7246, pp. 74–85. Springer, Heidelberg (2012). https://doi.org/10.1007/978-3-642-29066-4_7
15. Nobile, M.S., Cazzaniga, P., Besozzi, D., Pescini, D., Mauri, G.: cuTauLeaping: a GPU-powered tau-leaping stochastic simulator for massive parallel analyses of biological systems. PLoS ONE **9**(3), e91963 (2014)
16. Nvidia: CUDA C Best Practices Guide (2012)
17. Nvidia: Nvidia CUDA C Programming Guide 7.5 (2015)
18. Orellana, A., Minetti, G.F.: A modified binary-PSO for continuous optimization. In: XV Congreso Argentino de Ciencias de la Computación (2009)
19. Petre, I., et al.: A simple mass-action model for the eukaryotic heat shock response and its mathematical validation. Nat. Comput. **10**(1), 595–612 (2011)
20. Petzold, L.R.: Automatic selection of methods for solving stiff and nonstiff systems of ordinary differential equations. SIAM J. Sci. Stat. Comput. **4**, 136–148 (1983)
21. Szallasi, Z., Stelling, J., Periwal, V.: System Modeling in Cellular Biology: From Concepts to Nuts and Bolts. The MIT Press, Boston (2006)
22. Trelea, I.C.: The particle swarm optimization algorithm: convergence analysis and parameter selection. Inf. Process. Lett. **85**(6), 317–325 (2003)

23. Vitorino, L., Ribeiro, S., Bastos-Filho, C.J.: A mechanism based on artificial bee colony to generate diversity in particle swarm optimization. Neurocomputing **148**, 39–45 (2015)
24. Wolkenhauer, O., Ullah, M., Kolch, W., Kwang-Hyun, C.: Modeling and simulation of intracellular dynamics: choosing an appropriate framework. IEEE Trans. Nanobiosci. **3**(3), 200–207 (2004)

Haplotype and Repeat Separation
in Long Reads

German Tischler-Höhle[(✉)]

Myers Lab, Max Planck Institute of Molecular Cell Biology and Genetics,
Pfotenhauerstraße 108, Dresden, Germany
`tischler@mpi-cbg.de`
`https://myerslab.mpi-cbg.de/people/german-tischler/`

Abstract. Resolving the correct structure and succession of highly similar sequence stretches is one of the main open problems in genome assembly. For non haploid genomes this includes determining the sequences of the different haplotypes. For all but the smallest genomes it also involves separating different repeat instances. In this paper we discuss methods for resolving such problems in third generation long reads by classifying alignments between long reads according to whether they represent true or false read overlaps. The main problem in this context is the high error rate found in such reads, which greatly exceeds the variance between the similar regions we want to separate. Our methods can separate read classes stemming from regions with as little as 1% difference.

1 Scientific Background

Third generation sequencing reads like those produced by Pacific BioSciences (PacBIO) and Oxford Nanopore Technologies (ONT) sequencers are very long in comparison with the ones produced by second generation sequencers. The average read length for PacBIO is often 10 k base pairs (bp) and for ONT 7–16 kbp have been reported. For PacBIO more than half of the sequenced bases can be in reads of length 20 kbp and above. Second generation sequencers often yield reads as short as 150 bp, so third generation sequencers allow a much better repeat resolution for assembly because more repeats are spanned by single reads. The increased read length of third generation sequencers however comes at the price of a much higher average base error rate (about 13% for PacBIO and even higher for ONT).

This poses major algorithmic challenges in the areas of sequence alignment, comparison and signal detection. Read versus read comparison (see e.g. [1]) operates at a correlation of 70% and less. This makes it very hard to detect small differences between regions where reads were sampled from. Reads stemming from sufficiently similar regions in an underlying genome, like instances of a repeat or different haplotypes, will often align within the parameters used, as the difference between the two sources is small in comparison with the read error rate. Being able to segregate read alignments into classes according to whether

M. Bartoletti et al. (Eds.): CIBB 2017, LNBI 10834, pp. 103–114, 2019.
https://doi.org/10.1007/978-3-030-14160-8_11

or not an alignment between a read pair designates a real overlap in the underlying genome is however important for multiple applications like genome assembly and variant detection. In genome assembly for instance the quality of any consensus sequence produced rises and falls with the ability to select the correct reads as input (cf. [2]). Linking up different haplotypes during the assembly of a non haploid organism results in patchwork like output, in particular an assembly process yielding output contigs which are not contained in the genome to be reconstructed. Haplotype assembly designates the problem of separating reads into haplotype classes by first mapping them to a given reference sequence and then splitting the reads into groups using the information obtained. Several papers have presented methods for haplotype assembly (or read phasing) in the diploid setting (cf. [3–10]). Some of these methods are capable of employing base qualities to improve the phasing accuracy. The case of phasing in the more general polyploid setting has also been considered (see [11] and references therein). Canu (see [12]) performs repeat separation using a sequence of error correction, residual error estimation and classification of the error corrected reads. The authors report being able to separate repeat instances with 3% difference and above.

2   Materials and Methods

In this paper we discuss methods for repeat and haplotype separation in long reads. We consider the setting of de novo assembly, in particular we do not presume or require the existence of a reference sequence or known variation sites. Instead of read to reference alignments we use read to read consensus alignments, i.e. we align reads to error corrected reads. In addition we do not limit our attention to a scenario requiring there to be at most two versions of a sequence. Our setting is thus more general than the setting of haplotype assembly for diploid genomes.

We consider two basic principles for splitting a set of reads. The first one is based on the trivial observation that reads in the same class should agree on most positions, particularly including those for which a very rough analysis shows a potential for disagreement in the read set. As the reads we consider are not error free we cannot expect the reads inside one class to agree on all positions. This approach has its merits when the number of versions a sequence appears in is low but as we will see below, it becomes unsuitable as the number of versions grows. The second principle is based on observing sets of reads (more or less) consistently disagreeing on certain positions. This scales to higher version numbers but is computationally much more expensive.

2.1   Preliminaries

Let $G = \{S_1, S_2, \ldots, S_k\}$ denote a genome containing sequences S_i for $i = 1, \ldots, k$, i.e. strings over the alphabet $\Sigma = \{A, C, G, T\}$. Further let $R = \{R_1, R_2, \ldots R_r\}$ be a set of reads sampled from G (randomly of the forward and reverse complement strand) such that the strings in R have length L on average and the error rate (errors per length on G) between the reads and the intervals on G they were drawn from is p_e on average. For PacBIO the length distribution in R would follow a log normal distribution with average length 10kb and p_e would be in the order of 0.13.

We denote a local alignment between sequences U_i and U_j by a tuple $(i, j, ib, ie, jb, je, c)$ where ib and ie mark the start and end of the alignment on U_i, jb and je the start and end on U_j and c is a Boolean value marking whether U_j or the reverse complement of U_j was used ($c =$ true for reverse complement).

In practice long reads often contain stretches of very low quality, so even for two reads sharing a true overlap we find a sequence of local alignments instead of a single suffix/prefix or containment type alignment. Our methods can easily be generalised to this case, however for the sake of simplicity of exposition we assume that alignments between reads are contiguous below. Let A denote the set of all (local) alignments between pairs of reads in R s.t. the correlation between the two reads inside the alignment is at least $1 - 2p_e$ and the alignment covers at least ℓ bases on both reads involved for some length ℓ. In practice we commonly use $\ell = 1k$ for third generation long reads (which is the default setting for DALIGNER, see [1]). For a given read R_k we call the subset of A s.t. the first component of the tuples is k the alignment pile for R_k. If G represents a non haploid genome or contains sufficiently long repeating regions then not all the alignments in A may refer to true read overlaps on G.

We can use an alignment pile of a read or a subset thereof (for instance by choosing the top k best aligning other reads for some k) to compute a preliminary consensus or error corrected version of the read, e.g. using the algorithm proposed in [2]. We denote a preliminary consensus obtained for a read R_i in this way by $\hat{R}_i$. The alignment pile for R_i can be transformed into an alignment pile for $\hat{R}_i$ by aligning the reads in the pile for R_i to $\hat{R}_i$ while taking the positions of the original alignments on R_i into account and transforming those to positions on $\hat{R}_i$ using an alignment of R_i and $\hat{R}_i$. In addition to the original alignments in the pile of R_i we also insert an alignment between $\hat{R}_i$ and R_i into the pile of $\hat{R}_i$.

An alignment pile for some $\hat{R}_k$ can be transformed into a matrix where the columns represent positions on or before the bases of $\hat{R}_k$ (before for base insertions into $\hat{R}_k$) and the rows are the reads in the alignment pile for $\hat{R}_k$. An alignment $(k, j, kb, ke, jb, je, c)$ between $\hat{R}_k$ and a read j is active from base kb to ke on k. The cells of a matrix row corresponding to read R_j are set as follows. Columns the respective alignment is inactive on remain empty. In the active region of the alignment a cell is filled with the base from R_j if the alignment features a match, mismatch or insertion operation for the respective position and a dash ($-$) otherwise. As a convention we always have the alignment between $\hat{R}_k$ and R_k as the first row of the matrix. Table 1 shows an example. The excerpt shows positions 100 to 102 on the read. Some bases have been inserted before position 102, which is marked by the position identifier $(102, -1)$. The alignment corresponding to the second row ends at position 101, the one for the last row starts at position 101.

Table 1. Excerpt from a matrix given by an alignment pile

(100,0)	(101,0)	(102,-1)	(102,0)
A	C	A	T
A	C		
A	T	A	T
	C	–	G

2.2 Agreement Based Splitting

Let d denote the average sequencing depth of the read set R. We assume the arrival rate of reads on the genome follows a Poisson distribution with mean d, i.e. we have a probability of $P_d(i) = \frac{d^i}{i!}e^{-d}$ to see a depth of i at a given position. This describes the case of a uniform sampling of the underlying genome without any bias (e.g. an amplification bias common for sequencing methods requiring PCR based amplification).

The probability to see d' correctly sequenced bases for any position is thus

$$P_c(d') = \sum_{i=d'}^{\infty} P_d(i) \binom{i}{d'} (1 - p_e)^{d'} p_e^{i-d'} \tag{1}$$

We want to detect variation sites inside a given read R_i. One very simple way to do this is to scan the matrix constructed for $\hat{R}_i$ for columns in which more than one symbol appears with a frequency above a given threshold. Assuming the alignments used to construct the matrix are suitable we would see such a variation with probability $\sum_{j=d'}^{\infty} P_c(j)$ if we chose a threshold of d'. For $d = 20$ and $p_e = 0.15$ we obtain $d' = 8$ if we ask the probability to be at least 99%, i.e. we are 99% sure not to miss a relevant site if we look for columns containing at least two symbols with 8 or more instances. There is however the chance of calling variation sites because of unsuitable alignments in the pile for $\hat{R}_i$ or a sufficiently high number of wrongly sequenced bases (this is a problem especially in the presence of a high number of sequence versions as this increases the total number of reads involved in the pile).

Consider a given position q in the genome G and two reads R_i and R_j covering this position. Then we have a probability of $(1 - p_e)^2$ for having the base at position q sequenced correctly in both R_i and R_j. Let $A_{ij} = (i, j, ib, ie, jb, je, c)$ denote an alignment between read R_i and R_j and assume we called n variants on R_i inside the index interval $[ib, ie]$. If R_i and R_j overlap as designated by A_{ij} in the underlying genome, then we have a probability of

$$P_s(\kappa) = \sum_{\kappa'=\kappa}^{n} \binom{n}{\kappa'} (1 - p_e)^{2\kappa'} (1 - (1 - p_e)^2)^{n-\kappa'} \tag{2}$$

to see R_i and R_j agree on at least κ of the n disagreement points in the matrix for the alignment pile of $\hat{R}_i$.

If there are two underlying versions, e.g. a repeat with two copies or haplotypes in a diploid genome, then we would expect to see reads coming from different versions to disagree on most of the variant locations. In this case we have a strong signal for separating the two versions. It becomes weaker in the presence of more versions when some of the versions agree with others in a large fraction of the variant locations. In this case we cannot reliably tell the difference between two reads stemming from different versions with a relatively low number of sequencing errors and two reads stemming from the same version but agreeing on a lower number of variant locations due to a higher number of sequencing errors. For experiments we choose the number m of disagreement points two reads need to agree on so we consider them as from the same class as the smallest number s.t. $P_s(m) \geq 0.995$. The asymptotical time complexity of the algorithm for processing the pile of some read R is $O(\ell d)$ where ℓ denotes the length of R in bases and d is the number of alignments in the pile for R.

2.3   Disagreement Based Splitting

One of the main problems with agreement based splitting is suboptimal performance when reads from different classes agree on a large number of the detected variant locations. Splitting based on the differences between genomic regions does not suffer from this effect. Every attempt via directly comparing two long reads is however bound to fail as the high sequencing error rate drowns any slight difference between the two underlying real sequences. At a single base error rate of $p_e = 15\%$ the probability to see a correct pair of corresponding bases in two reads is $(1 - p_e)^2 = 72.25\%$, i.e. 27.75% of the pairs are wrong and most of these wrong pairs lead to a false disagreement between reads which should agree.

When we compare bases for discovering disagreements between reads, we need to make reasonably sure that the bases compared are correct representations of their class for a given position. Consider some position on k reads stemming from the same class. Then we have a probability of $1 - p_e^k$ to see the correct base in at least one of these k reads. For $k = 2$ and $p_e = 0.15$ we have a probability of 97.75%, still a probability of more than 2% for all the bases to be wrong, for $k = 3$ we reach 99.6625%.

In consequence, if three reads from the same class agree on a base, then this is most likely a correctly reported base. We use this observation by comparing three tuples of bases to three tuples of bases instead of comparing single read bases to single read bases.

Given a read R_j we first build the matrix corresponding to the alignment pile of $\hat{R}_j$. We then scan the matrix column for column. In each column c we extract all 6 tuples $(r_1, r_2, r_3, r_4, r_5, r_6)$ s.t. r_i for $1, 2, \ldots, 6$ are row identifiers marking non empty cells in column c, the cells for row r_1, r_2 and r_3 all contain the same symbol a, the cells for row r_4, r_5 and r_6 all contain the same symbol b, $a \neq b$, $1 = r_1 < r_2 < r_3$ and $r_4 < r_5 < r_6$. Remember row 1 in the matrix refers to the alignment between $\hat{R}_j$ and R_j. There are $O\left(\binom{q}{2}\binom{q}{3}\right) = O(q^5)$ distinct such tuples in the worst case if q is the maximum number of active alignments in any column of the matrix. For each distinct tuple T we count the number $Y(T)$ of times it appears summed up over all columns. The support $Z(T)$ of a tuple T is the intersection of the active intervals of the alignments it is based on.

If we want to split read sets down to a difference rate of δ, then we expect $\Delta = \delta|Z(T)|$ differences to exist inside $Z(T)$. Assuming suitable alignments comprising R_j's matrix, the probability to see each single of these differences is $p_6 = (1-p_e)^6$ which is about 37.7% for $p_e = 15\%$. The probability to see at least m of these differences is $\eta(i) = \sum_{i=m}^{\Delta} \binom{\Delta}{i} p_6^i (1-p_6)^{\Delta-i}$. We choose the smallest i s.t. $\eta(i)$ is at least 99.5% as a threshold. For each i we count the number $H(i)$ of tuples satisfying their threshold in which i appears as r_4, r_5 or r_6.

Given P_d (Poisson distribution) as defined above we can determine a depth threshold d_t which is reached for most bases on the genome. Using the average sequence depth we can also estimate the likelihood of having a certain number v of sequence variants in the pile observed. Reads i with a count $H(i)$ close to or exceeding $h_t = \binom{d_t}{2}\binom{(v-1)d_t}{2}$ (we have fixed r_1 to 1 and one of r_4, r_5 or r_6 to r) are most likely not in the same class at R_j. Reads i in the same class as R_j should have a $H(i)$ equal or close to zero.

The asymptotical time complexity of the algorithm for processing the pile for some read R containing d alignments is dominated by enumerating the 5-tuples, which is $O(\ell d^5)$ where ℓ denotes the length of R in bases. Note that we do not need an array of size d^5 for counting the frequency of each distinct tuple. The tuples originating from each single matrix column for a read R_j can easily be enumerated in lexicographically increasing order in the following way. Let $\alpha = \alpha_1, \alpha_2, \ldots, \alpha_y$ be the sequence of read ids in the column featuring the same symbol as R_j in increasing order excluding R_j itself and $\beta = \beta_1, \beta_2, \ldots, \beta_z$ the sequence of read ids in the column featuring a different symbol from R_j's in increasing order. If $y < 2$ or $z < 3$ then there are no tuples for this column. Otherwise we keep two pointers a_1 and a_2 into α s.t. $1 \leq a_1 < a_2 \leq y$ starting with $a_1 = 1, a_2 = 2$ and three pointers b_1, b_2 and b_3 into β s.t. $1 \leq b_1 < b_2 < b_3 \leq z$ starting with $b_1 = 1, b_2 = 2, b_3 = 3$. These pointers encode a tuple $(\alpha_{a_1}, \alpha_{a_2}, \beta_{b_1}, \beta_{b_2}, \beta_{b_3})$. The next tuple is obtained by updating the pointers. If $b_3 < z$ then we increment b_3. Otherwise if $b_2 + 1 < z$ then we increment b_2 and then set b_3 to $b_2 + 1$. Otherwise if $b_1 + 2 < z$ then we increment b_1 and then set b_2 to $b_1 + 1$ and b_3 to $b_1 + 2$. Otherwise if $a_2 < y$ then we increment a_2 and set $b_1 = 1, b_2 = 2, b_3 = 3$. Otherwise if $a_1 + 1 < y$ then we increment a_1 set a_2 to $a_1 + 1$ and $b_1 = 1, b_2, b_3$. Otherwise there is no next tuple. It is thus sufficient to have the arrays α and β as well as 5 pointers to enumerate the tuples originating from a column in lexicographically increasing order.

For enumerating all tuples originating from the matrix in lexicographically increasing order we use a min heap containing the current tuple for each column encoded in pointer form where comparisons between heap elements are performed using decoded tuples from those pointers.

Let ℓ denote the length of R_j in bases and d the number of alignments in it's alignment pile. Concerning space we require the arrays α and β as well as the 5 pointers for each column plus a heap containing a maximum of ℓ pointer sets. This space usage is in $O(\ell d)$.

3   Results

We have implemented both splitting methods. They are freely available as the programs split_agr and split_dis in the daccord package (see https://gitlab.com/german.tischler/daccord). The daccord program (see [2]) in this package was also used to compute preliminary consensus sequences for the splitting. Read versus read alignments were computed using DALIGNER (cf. [1]).

We performed two types of performance tests, both of which are based on simulated reads to ensure we can properly check whether and to what degree splittings computed are accurate. For the sake of comparison with other state of the art solutions we compared our results to those computed by WhatsHap (cf. [4]) for the diploid setting. As far as polyploid phasing is concerned we are not aware of another practical implementation solving the problem we discuss in this paper, i.e. de novo detection of variants from long noisy reads only and subsequently assigning haplotypes/repeat ids to these reads. All tests were run on 24 core Intel Xeon E5-2680 v3 systems equipped with 256G of RAM.

In the first test we took a 190kb piece of the E. coli genome, duplicated it k times for $k = 1, 2, \ldots 7$ and spiked in 1% difference between the duplicated versions and the original. The differences are single bp insertions, deletions and substitutions with equal probability.

We generated reads of average length 15 kbp to evenly cover the sequences at depth $d = 20$. The reads produced have stretches of low average error rate drawn from a normal distribution with average 15% and standard deviation 3% as well as stretches of high error rate drawn from a normal distribution with an average of 40% and standard deviation 3%. The generator uses a simple two state probabilistic finite state machine to determine whether to stay in a low or high error region or whether two switch to the other. The probabilities used are 0.9998 to stay in low error mode and 0.995 to stay in high error mode, while we start in low error mode with a probability of 0.7. Thus the error rate can greatly vary along a read. Of the errors inserted 80% were insertions, 13.3% deletions and 6.6% substitutions (see [13]). An overview over available read simulators is given in [14]. None of the published long read simulators (e.g. [15]) met our requirement of producing even coverage while generating reads following a log normal length distribution and with a wide range of different error rates a described above (as it is commonly found in long reads), so we used our own simulator to generate synthetic reads (available as the longreadgen program in the loresim2 package, see https://gitlab.com/german.tischler/loresim2). One crucial feature missing from most simulators is the generation of low quality strechtes inside reads. Such regions are quite common in real read data and cause partial instead of complete alignments between reads. Without this feature we think the reads simulated would be rather unrealistic.

For the splitting we only considered read overlaps of 5 kbp and more to reduce noise in the underlying statistics. The benchmarking scripts used to produce the data shown below can be found at https://gitlab.com/german.tischler/daccord_separation_benchmarks.

Table 2. Performance of splitting on 190 kbp stretch of E. coli with $1-7$ copies added at 1% difference to original

Agreement based					Disagreement based				
Copies	Precision	Recall	F_1	t	Copies	Precision	Recall	F_1	t
1	0.999	0.788	0.881	1.48	1	0.999	0.981	0.990	4.52
2	0.980	0.634	0.770	2.31	2	0.999	0.949	0.973	32.4
3	0.904	0.721	0.802	3.17	3	0.999	0.977	0.988	139
4	0.719	0.829	0.770	4.26	4	0.998	0.983	0.991	411
5	0.443	0.910	0.596	5.96	5	0.998	0.986	0.992	1055
6	0.218	0.957	0.355	8.17	6	0.998	0.986	0.992	2383
7	0.139	0.981	0.244	10.76	7	0.997	0.980	0.989	4784

Table 2 shows the performance of agreement and disagreement based splitting in this scenario. We provide precision (which fraction of the alignments kept is true), recall (which fraction of the true alignments is kept) and F_1 (harmonic mean of precision and recall) score measures to quantify the performance of the read classification methods. The last column provides the average run-time of the splitting programs in seconds per read. All scores given are rounded to 3 significant decimals. For agreement based splitting the performance values in the respective tables only include data for such reads where the method is able to detect variants for at least 0.1% (i.e. 1 in 1000) of the read bases. Reads below this threshold are marked as unsplit in the output.

While agreement based splitting has good precision for one and two modified copies, the performance quickly drops to the point where essentially most wrong alignments are kept. The disagreement based splitting works very well in this setting. For computing the threshold h_t we have provided the correct value for the number of variants v to the program, as it does not yet support estimating it from the input data. For the separation of read piles caused by repeats this parameter could in principle be estimated from the read data using an estimate of the sequencing depth obtained by inspecting the pairwise alignments and counting the number of reads aligned to each single base of a read. The ploidy of a sequenced organism is probably most easily observed by wet lab based methods. The δ parameter was set to 1%.

While the run time grows essentially linearly with the number of copies for agreement based splitting, the growth is a lot steeper for the disagreement based method due to the factor of $(dv)^5$ (where we keep $d = 20$ constant in our experiments, so the growth is with v^5).

As the first test is highly synthetic, we have chosen a somewhat more realistic scenario for the second one. We have extracted regions containing the genes FCGR1(A|B|CP), FCGR2(A|B|C) and FCGR3(A|B) plus 100 kbp to the left and right of these regions from chromosome 1 of the human reference genome (GRCh38). These regions are highly repetitive with repeating stretches of length up to 46 kbp with a difference of merely 1% and one repeat of length 26 kbp with 0.4% difference between the copies.

Table 3. Performance of splitting on FCGR regions of human chromosome 1

Agreement based				Disagreement based			
Precision	Recall	F_1	t	Precision	Recall	F_1	t
0.933	0.824	0.875	1.36	0.925	0.947	0.936	6.03

We generated reads and alignments using the same parameters as for the other test. Table 3 shows the performance of the splitting approaches we measured. While the region considered is repetitive in its entirety, we do not have many cases of stretches appearing more than twice in total, i.e. most repeats have only two instances. As this is the setting in which agreement based splitting mostly works, we see a decent performance for this method in terms of precision as reflected in the table. In comparison the disagreement based splitting, for which we used $\delta = 1\%$ and $v = 2$, yields a slightly lower precision value while the recall value is superior. A closer look reveals that the average difference between the true sequences we fail to separate (which lead to the false positive alignments we keep) is about 0.5% which is way below our setting for δ, so the failure in separation is not surprising. Just reducing the parameter δ below 1% however does not markedly improve the splitting, as this also greatly increases noise (disagreement tuples observed although they are not real). Improving this provides opportunities for additional research.

For the sake of putting our results into perspective relative to other state of the art software we in the following provide some performance scores reached by WhatsHap (cf. [4]) on the diploid (like) settings supported by that program. This concerns the E.coli test with one added modified copy at 1% difference and the FCGR region test.

WhatsHap requires alignments in the form of a BAM file and a VCF files containing variant sites as input. To provide these we first computed a corrected version $\hat{R}$ for each read R using daccord (see [2]) and then transformed all alignments (R, Q) with another read Q produced by DALIGNER into alignments $(\hat{R}, Q)$ and produced a BAM file containing these alignments. The corrected version $\hat{R}$ was used to provide a situation similar to the alignment of sequencing reads to a error free/low error reference sequence as it is common for BAM files. We then used FreeBayes [16] as suggested by the WhatsHap documentation to produce a VCF file containing variants. These two files were then used as input for WhatsHap to first compute a phased VCF file (using WhatsHap's phase command) and then a coloured BAM file (using WhatsHap's hapltotag command). For score values we considered only such reads R in the coloured BAM file where the alignment $(\hat{R}, R)$ had been coloured. For those reads we considered alignments $(\hat{R}, Q)$ as correctly coloured if $(\hat{R}, Q)$ was assigned the same haplotype as $(\hat{R}, R)$ and (R, Q) is a true overlap or $(\hat{R}, Q)$ was assigned a different haplotype as $(\hat{R}, R)$ and (R, Q) is not a true overlap.

Table 4. Performance of WhatsHap colouring on E.coli test and FCGR regions using FreeBayes variant calls

E.coli				FCGR			
Precision	Recall	F_1	t	Precision	Recall	F_1	t
0.488	0.991	0.654	5.46	0.405	0.993	0.576	7.15

Table 4 shows the performance values achieved. The time given only refers to the time used by the WhatsHap program, the variant calling by FreeBayes is not included. In both cases the performance in terms of the precision score is quite bad, less than half of the alignments retained for a read are true overlaps. This result left us wondering whether the poor performance is due to inadequate variant detection by FreeBayes (considering the fact that it is described as a short read variant detector) or a poor use of the information provided by WhatsHap. Attempts at different ways to produce a suitable VCF file using standard tools failed (samtools mpileup crashed on the BAM files provided, GATK Haplotype-Caller produced essentially empty VCF files). As a last resort we tried converting the set of variants detected by our agreement based splitting to the VCF format, which gave us the performance values shown in Table 5. For the E.coli test based on simple SNP and single base insertion and deletions the colouring by WhatsHap is close to perfect, although it does not quite reach the precision score of our methods presented in this paper. For the FCGR test based on partly more complex variations the performance of WhatsHap is still a lot worse than the new solutions presented in this paper. The times provided are those required for the WhatsHap runs plus the times required for the respective runs of the agreement based splitting used for calling the variants.

Table 5. Performance of WhatsHap colouring on E.coli test and FCGR regions using agreement based variant calling as proposed in this paper

E.coli				FCGR			
Precision	Recall	F_1	t	Precision	Recall	F_1	t
0.998	1	0.999	3.88	0.642	0.944	0.764	4.48

4   Conclusion

We have shown that repeat and haplotype separation in long reads with current read length and error rates is possible down to a difference of 1% and possibly less. If our preliminary results extend to more general settings, then this may improve on the current state of the art of 3% set by Canu. The methods proposed also work if there are more than two underlying sequence versions. We hope these new insights can help to significantly improve the assembly of repetitive regions in genomes. The disagreement based splitting is very computationally expensive in its pure and exhaustive form. We are however hopeful that a randomised approach will yield similar performance values. To this end we will randomly subsample alignments inside a pile and perform the disagreement based splitting on such piles. The repeated application of such a subsampling experiment will improve the splitting similar to the Miller-Rabin randomised primality test.

Acknowledgments. We thank Gene Myers for interesting algorithmical discussions related to this paper and Shilpa Garg for advice on running WhatsHap.

References

1. Myers, G.: Efficient local alignment discovery amongst noisy long reads. In: Brown, D., Morgenstern, B. (eds.) WABI 2014. LNCS, vol. 8701, pp. 52–67. Springer, Heidelberg (2014). https://doi.org/10.1007/978-3-662-44753-6_5
2. Tischler, G., Myers, E.W.: Non hybrid long read consensus using local de bruijn graph assembly. bioRxiv (2017). https://www.biorxiv.org/content/early/2017/02/06/106252
3. Patterson, M., Marschall, T., Pisanti, N., van Iersel, L., Stougie, L., Klau, G.W., Schönhuth, A.: WHATSHAP: haplotype assembly for future-generation sequencing reads. In: Sharan, R. (ed.) RECOMB 2014. LNCS, vol. 8394, pp. 237–249. Springer, Cham (2014). https://doi.org/10.1007/978-3-319-05269-4_19
4. Murray, P., et al.: WHATSHAP: weighted haplotype assembly for future-generation sequencing reads. J. Comput. Biol. **22**(6), 498–509 (2015). https://doi.org/10.1089/cmb.2014.0157. pMID: 25658651
5. Martin, M., et al.: WHATSHAP: fast and accurate read-based phasing. bioRxiv (2016). https://www.biorxiv.org/content/early/2016/11/14/085050
6. Bansal, V., Halpern, A.L., Axelrod, N., Bafna, V.: An MCMC algorithm for haplotype assembly from whole-genome sequence data. Genome Res. **18**(8), 1336–1346 (2008). http://genome.cshlp.org/content/18/8/1336.abstract
7. Bansal, V., Bafna, V.: Hapcut: an efficient and accurate algorithm for the haplotype assembly problem. Bioinformatics **24**(16), i153–i159 (2008). https://doi.org/10.1093/bioinformatics/btn298
8. Mazrouee, S., Wang, W.: Fasthap: fast and accurate single individual haplotype reconstruction using fuzzy conflict graphs. Bioinformatics **30**(17), i371–i378 (2014). btu442[PII], http://www.ncbi.nlm.nih.gov/pmc/articles/PMC4147895/
9. Deng, F., Cui, W., Wang, L.: A highly accurate heuristic algorithm for the haplotype assembly problem. BMC Genomics **14**(2), S2 (2013). https://doi.org/10.1186/1471-2164-14-S2-S2
10. Chin, C.S., et al.: Phased diploid genome assembly with single-molecule real-time sequencing. Nat. Methods **13**, 1050 EP (2016). https://doi.org/10.1038/nmeth.4035. article
11. Chaisson, M.J., Mukherjee, S., Kannan, S., Eichler, E.E.: Resolving multicopy duplications *de novo* using polyploid phasing. In: Sahinalp, S.C. (ed.) RECOMB 2017. LNCS, vol. 10229, pp. 117–133. Springer, Cham (2017). https://doi.org/10.1007/978-3-319-56970-3_8
12. Koren, S., Walenz, B.P., Berlin, K., Miller, J.R., Bergman, N.H., Phillippy, A.M.: Canu: scalable and accurate long-read assembly via adaptive k-mer weighting and repeat separation. Genome Res. **27**(5), 722–736 (2017). http://genome.cshlp.org/content/27/5/722.abstract

13. Carneiro, M.O., Russ, C., Ross, M.G., Gabriel, S.B., Nusbaum, C., DePristo, M.A.: Pacific biosciences sequencing technology for genotyping and variation discovery in human data. BMC Genomics **13**(1), 375 (2012). https://doi.org/10.1186/1471-2164-13-375

14. Escalona, M., Rocha, S., Posada, D.: A comparison of tools for the simulation of genomic next-generation sequencing data. Nat. Rev. Genet. **17**(8), 459–469 (2016). 27320129[pmid]. http://www.ncbi.nlm.nih.gov/pmc/articles/PMC5224698/

15. Ono, Y., Asai, K., Hamada, M.: Pbsim: Pacbio reads simulator-toward accurate genome assembly. Bioinformatics **29**(1), 119–121 (2013). https://doi.org/10.1093/bioinformatics/bts649

16. Garrison, E., Marth, G.: Haplotype-based variant detection from short-read sequencing. ArXiv e-prints (2012)

Tumor Subclonal Progression Model for Cancer Hallmark Acquisition

Yusuke Matsui[1(✉)], Satoru Miyano[2], and Teppei Shimamura[3]

[1] Graduate School of Medicine, Laboratory of Intelligence Healthcare,
Nagoya University, Nagoya, Japan
ymatsui@med.nagoya-u.ac.jp
[2] Institute of Medical Science, Laboratory of DNA Information Analysis,
Human Genome Center, The University of Tokyo, Tokyo, Japan
[3] Graduate School of Medicine, Division of Systems Biology,
Nagoya University, Nagoya, Japan

Abstract. Recent advances in the methods for reconstruction of cancer evolutionary trajectories opened up the prospects of deciphering the subclonal populations and their evolutionary architectures within cancer ecosystems. An important challenge of the cancer evolution studies is how to connect genetic aberrations in subclones to a clinically interpretable and actionable target in the subclones for individual patients. In this study, our aim is to develop a novel method for constructing a model of tumor subclonal progression in terms of cancer hallmark acquisition using multiregional sequencing data. We prepare a subclonal evolutionary tree inferred from variant allele frequencies and estimate pathway alteration probabilities from large-scale cohort genomic data. We then construct an evolutionary tree of pathway alterations that takes into account selectivity of pathway alterations via selectivity score. We show the effectiveness of our method on a dataset of clear cell renal cell carcinomas.

Keywords: Cancer evolution · Multiregional sequencing ·
Pathway alteration · Clear cell renal cell carcinomas

1 Scientific Background

Cancer is a heterogeneous genetic disease characterized by dynamic evolution through acquisition of genomic aberrations. The clonal theory of cancer proposed by Nowell [1] postulates that acquisition of a mutation in cancer follows natural selection of the Darwinian model, in which cancer obtains the advantages of biological fitness under selective pressure.

The development of multiregional sequencing techniques has provided new perspectives on genetic heterogeneity [2]. According to studies on multiregional sequencing, spatially distinct regions within the same tumor acquire different sets of somatic single-nucleotide variants (SSNVs), and this phenomenon is called

M. Bartoletti et al. (Eds.): CIBB 2017, LNBI 10834, pp. 115–123, 2019.
https://doi.org/10.1007/978-3-030-14160-8_12

intratumor heterogeneity. Recently, methods for reconstruction of cancer evolutionary structures were extensively studied. Because the cell population of each region is a mixture of normal and tumor cells, distinct regions are deconvoluted into cell subpopulations called subclones, and then they are assigned to tree structures under the constraint that is derived from the infinite site assumption [3].

Nonetheless, identification of clinically actionable subclone targets for individual patients remains problematic. One of the reasons is that resulting subclonal trees are too diverse to interpret with clinical information. In this direction, Matsui *et al.* [4] proposed clustering methods for cancer evolutionary trees based on the tree shapes and interpreting clinical impact of subclonal evolution via clustering results with clinical information of each tree. Another reason is the difficulties with identifying the most plausible biological event from the limited number of SSNVs because of the sequencing depth and low frequencies of mutations.

With the aim to overcome this problem, we developed a novel method for inferring a tumor progression model in terms of acquisition of cancer hallmarks (Fig. 1). Our contributions are as follows: (1) proposing a novel framework for interpreting the cancer evolutionary tree using pathways and cancer hallmarks (2) developing the integrative approach to estimate individual each subclone's cancer hallmarks by merging multiregional sequencing data and large scale genomic cohort data. We also demonstrate the effectiveness of our method on an actual dataset of clear cell renal cell carcinomas (ccRCCs [7]).

2   Materials and Methods

Our method consists of 3 steps: (1) constructing a skeleton of a cancer subclonal evolutionary tree, (2) estimation of pathway alteration probability and calculation of *selectivity score*, and (3) constructing a model of progression of pathway alterations.

First, we construct an a priori evolutionary tree of the pathway alteration progression model, called a skeleton, to decompose the cell population into subclones and infer the subclonal evolutionary structures for each patient on the basis of multiregional variant allele frequencies (VAFs). Second, we estimate the pathway alteration probability by means of a large cohort dataset to identify the most likely pathway alterations in the subclones and to calculate *selectivity score*, *i.e.*, the strength of the selectivity among the pathway alterations at the subsequent step. At the last step, we construct the tumor progression model of pathway alterations based on the skeleton and pathway alteration probabilities under three assumptions as follows:

(Assumption 1) No pathway alteration occurs twice in the course of cancer evolution.

(Assumption 2) No pathway alteration is ever lost.

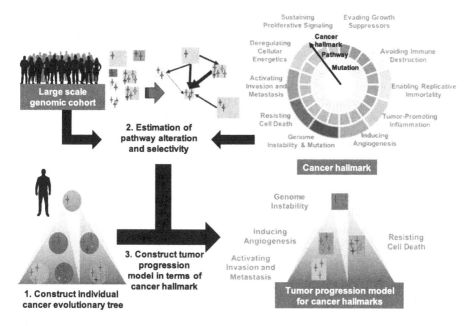

Fig. 1. Overview of our proposed approach for inferring tumor progression model for cancer hallmarks to interpret the biological functions of each subclones. The proposed method consists of 3 steps: (1) constructing a cancer subclonal evolutionary trees for individuals by multiregional sequencing data, (2) estimating pathway alteration probabilities and the strength of selective pressure between pathway alterations, and (3) constructing a model of progression of pathway alterations with cancer hallmarks.

Assumptions 1 and 2 mean that if a given pathway alteration occurs in any subclone, it happens exactly once in the course of tumor progression. We reconstruct the progression model from ancestral subclones, and we never use the pathway alterations in those subclones for their descendant subclones. In addition to the 2 assumptions, we assume selective pressures between the pathway alterations.

(**Assumption 3**) There is selective pressure between the pathway alterations.

We model this system on the basis of the notion of conditional probability, which can be estimated from large-scale cohort datasets. Using the 3 assumptions, we identify a unique pathway alteration from the multiple candidates of pathway alterations with the strongest selectivity. In the following section, we describe our method in more detail.

2.1 Constructing a Skeleton of the Cancer Subclonal Evolutionary Tree

This skeleton represents the subclonal evolutionary tree based on VAFs obtained by bulk sequencing from a single patient with multiple regions. The VAFs are approximately proportional to the sizes of cell populations with the set of SSNVs; however, in the settings of bulk data, each region may be a mixture of normal and tumor cells and may require deconvoluting the cell populations into subpopulations. The identified subclones are assigned to tree structures via 2 assumptions: (i) a mutation cannot recur in the course of cancer evolution, and (ii) no mutation can be lost [6]. Several approaches are implemented to deal with the problem. Using one of the algorithms, LICHeE [3], we construct the skeleton from the VAFs for each patient.

Let $T_i^0 = (V_i, E_i)$ be a skeleton of patient $i; i = 1, 2, \ldots, n$ with a set of vertices $V_i = \{v_{ij}; i = 1, 2, \ldots, n, j = 1, 2, \ldots, \eta_i\}$ and edges $E_i = \{e_{ik}; i = 1, 2, \ldots, p, k = 1, 2, \ldots \nu_i\}$, where vertices and edges represent subclones with a set of SSNVs and evolutionary relations, respectively. Without a loss of generality, $v_{i,j=1}$ always represents normal cells. Each vertex has a set of labels that can be obtained from the mapping $L : v_{i,j} \mapsto L_{i,j}$, e.g., $L_{i,j} = \{\text{SSNV1, SSNV2, SSNV3}\}$.

2.2   Estimation of Pathway Alteration Probability and Selectivity Score

To carry out phenotypic characterization of each subclone, we need to identify the most closely related pathway alterations for subclones. There are mainly 2 approaches to detection of pathway alterations: one is knowledge-based gene enrichment analysis such as Fisher's exact test, and the other is a *de novo*–oriented approach, where the alteration patterns are mapped to large-scale protein networks and identify subnetworks as a driver pathway with cost functions such as a tendency for mutual exclusivity. In this study, we focus on the knowledge-based approach because the biological validation for *de novo* pathways is usually difficult to perform quickly.

Using the large-scale cohort data, because of the nature of the limitation on sample size in experimental data, we estimate the pathway alteration probability using SLAPenrich [5], which is a state-of-the-art method for identifying pathway alteration and provides background pathway alteration probabilities and mutual exclusivity. Let P denote the pathway list, and then the main output of SLAPenrich is the P-value $p_\xi; \xi = 1, 2, \ldots, |P|$ for each pathway ξ that represents the significance of the enrichment of mutations. Optionally, we can obtain pathway alteration probabilities for each sample, that is, $\rho_{\xi,\zeta} = \Pr(X_{\xi,\zeta} \geq 1); \xi = 1, 2, \ldots, |P|, \zeta = 1, 2, \ldots, N$ where $X_{\xi,\zeta}$ represents the number of SSNVs that are included in pathway ξ in sample ζ.

We define binary variable $y_{\xi,j}$ if $\rho_{\xi,\zeta} > t$ then $y_{\xi,\zeta} = 1$ otherwise $y_{\xi,\zeta} = 0$ with a given threshold $0 \leq t \leq 1$. We evaluate the selectivity score, $S_{\alpha \to \beta}; \alpha, \beta = 1, 2, \ldots, |P|, \alpha \neq \beta$, defined as

$$S_{k \to l} = \Pr(Y_\alpha | Y_\beta = 1) - \Pr(Y_\alpha | Y_\beta = 0) \tag{1}$$

where Y_α represents a variable denoting the alteration status of pathway α. We empirically estimate $S_{\alpha \to \beta}$ as follows:

$$\tilde{S}_{\alpha \to \beta} = \frac{\sum_{\zeta=1}^{N} y_{\alpha,\zeta} | y_{\beta,\zeta}=1}{\sum_{\zeta=1}^{N} y_{\beta,\zeta}} - \frac{\sum_{\zeta=1}^{N} y_{\alpha,\zeta} | y_{\beta,\zeta}=0}{\sum_{\zeta=1}^{N} (1 - y_{\beta,\zeta})}. \tag{2}$$

We consider zero or negative values to represent a no selectivity, i.e., $\tilde{S}_{\alpha \to \beta} = 0$ if $\tilde{S}_{\alpha \to \beta} \leq 0$.

2.3   Constructing a Model of Progression of Pathway Alterations

Now we are ready to construct the subclonal evolutionary tree for pathway alterations. Given a subclone, we first scan the SSNVs to identify candidate pathway alterations. If at least 1 SSNV is included in a pathway, this situation is called a *candidate pathway alteration*. Suppose $P_\xi; \xi = 1, 2, \ldots, |P|$ is the genes included in pathway ξ and $Z_{i,j,\xi}$ is the candidate pathway alteration status of subclone j in patient i where $Z_{i,j,\xi} = 1$ if $L(v_{i,j}) \subseteq P_\xi; i = 1, 2, \ldots, n, j = 1, 2, \ldots, \eta_i$ otherwise $Z_{i,j,\xi} = 0$. In case of $Z_{i,j,\xi} = 0$ for all ξ, we regard the subclone as a nonfunctional one and remove node $v_{i,j}$ and the corresponding edges from skeleton T_i^0.

By means of candidate pathway alterations $Z_{i,j,\xi}$, we identify the unique pathway alteration. In the event that an ancestral subclone consists of normal cells, we select a pathway with the smallest P-value, i.e.,

$$\operatorname{argmin}_\xi p_\xi \text{ for } Z_{i,j,\xi} = 1 \text{ and } \xi \notin Q \tag{3}$$

Otherwise, we select the pathway with the highest selectivity score, i.e.,

$$\operatorname{argmax}_\beta S_{\alpha \to \beta} \text{ for } Z_{i,j,\xi} = 1 \text{ and } \xi \notin Q \tag{4}$$

where α is the pathway alteration of the ancestral subclone, and Q is a set of pathway alterations that have already appeared in the ancestral subclones. Assumptions (1) and (2) are ensured by the condition $\xi \notin Q$. If there is no corresponding pathway alteration because all the candidate pathway alterations have already taken place in the ancestral subclones, then we remove the subclone as a nonfunctional one.

2.4   The Dataset

A dataset from a study on ccRCCs [7] was used for the analysis. Whole-exome multiregional bulk sequencing was performed on 8 individuals with clinical information, and 587 out of 602 mutations remained after filtering of mutations with depth less than $100\times$.

The estimation of pathway alteration probabilities followed SLAPenrich procedures described in [5]. The 417 KIRC (corresponding to ccRCC) samples from

The Cancer Genome Atlas (TCGA) and International Cancer Genome Consortium (ICGC) and high-confidence variants identified in another study, [8], were used for estimation of the pathway alteration probabilities. Pathway gene sets were downloaded from the Pathway Commons data portal (v8, 2016/04), and gene sets containing fewer than 4 or more than 1,000 genes were discarded. After merging the gene sets that correspond to the same pathway across multiple data sources or have a large overlap defined as Jaccard index ≥ 0.8, we obtained 1,911 pathway gene sets. Cancer hallmarks were assigned to 456 pathways [5].

3   Results

We reconstructed the skeletons from VAFs by means of LICHeE using the same parameters in their experimental settings that are described in [3], and eventually, 8 skeletons were obtained. Next, we estimated the pathway alteration probabilities based on SLAPenrich and obtained P-values for 209 pathways and pathway alteration probabilities for 417 patients. We evaluated the selectivity score with the threshold $t = 0.1$.

We show the results of the constructed models of ccRCC progression in terms of cancer hallmark acquisition in Fig. 2. The complexity of cancer evolutionary trees were reduced because several SSNVs in subclones plays the similar roles in terms of the biological pathway. Specifically, EV003 were reduced to only the two pathways alteration, which means most of SSNVs in the subclones are involved in the TP53 related pathway. In this way, our approach could summarize the original cancer evolutionary trees and give biological interpretation via mapping SSNVs to the biological pathways.

Table 1. Counts of subclones (patients) with each cancer hallmark. For example, "10 (7)" in Sustaining Proliferative Signaling means that 10 subclones have the cancer hallmark, and it was observed in 7 patients. Columns "trunk" and "private" list the numbers of cancer hallmarks in the common ancestral subclone and in subclones without any descendants, respectively.

Cancer hallmarks	Total number	Trunk	Private	Other
Sustaining Proliferative Signaling	10 (7)	1 (1)	5 (5)	4 (1)
Evading Growth Suppressors	6 (4)	0 (0)	2 (2)	4 (2)
Avoiding Immune Destruction	1 (1)	0 (0)	1 (1)	0 (0)
Enabling Replicative Immortality	2 (2)	0 (0)	1 (1)	1 (1)
Tumor-Promoting Inflammation	2 (2)	0 (0)	2 (2)	0 (0)
Activating Invasion and Metastasis	7 (5)	0 (0)	6 (5)	1 (0)
Inducing Angiogenesis	12 (6)	4 (4)	6 (4)	2 (1)
Genome Instability and Mutation	4 (3)	2 (2)	2 (1)	0 (1)
Resisting Cell Death	4 (4)	0 (0)	3 (3)	1 (1)
Deregulating Cellular Energetics	3 (3)	0 (0)	0 (0)	3 (3)

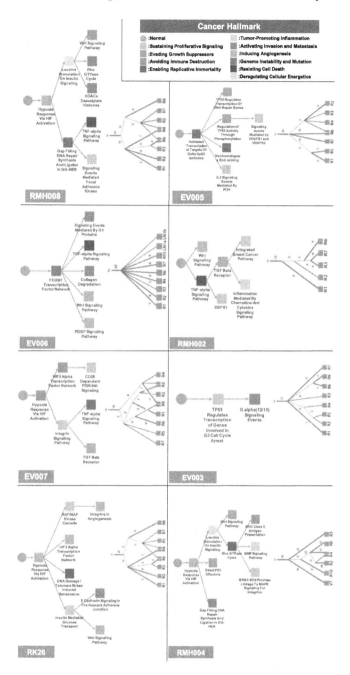

Fig. 2. ccRCC progression models of acquisition of cancer hallmarks (left side in the panels) with skeletons from LICHeE (right side in the panels and the figure is the same in [3]). The circle and square shapes indicate the normal cell population and subclone population in the proposed progression model, respectively. Cancer hallmarks are represented as colors shown in the top left panel. Pathway names are described below the box.

Next, we count the cancer hallmarks observed in the common ancestral sub-clones (trunk) and subclones without any descendants (private) as shown in Table 1.

The patterns of cancer hallmark acquisition still seemed diverse among patients; however, there were several patterns related to phenotypes when we focused on the trunk and private subsets. In the trunk subset, the "Inducing Angiogenesis" (4 subclones were counted) was the most frequently observed cancer hallmark that was due to pathway alterations caused by VHL mutations. The second most frequently observed cancer hallmark was "Genome Instabil-ity and Mutation" (2) caused by transcription factor–related aberrations such as the FOXM1 Transcription Factor Network. In the private subset, Sustaining Proliferative Signaling (5) and Activating Invasion and Metastasis (6) were the most common events among the patients (5 out of 8 patients). In particular, untreated patients (RMH004, RMH008, and RK26) showed Activating Invasion and Metastasis (6).

We also determined the frequency of evolutionary paths of cancer hallmarks up to 2 descendants. The most frequent path is "Normal - Inducing Angio-genesis - Deregulating Cellular Energetics" (3) and the second most frequent paths are "Normal - Inducing Angiogenesis - Inducing Angiogenesis" (2), "Nor-mal - Genome Instability and Mutation - Inducing Angiogenesis" (2), "Normal - Genome Instability and Mutation - Activating Invasion and Metastasis" (2), and "Normal - Genome Instability and Mutation - Evading Growth Suppres-sors" (2). These results give us biological and clinical implications beyond the SSNVs.

4   Conclusion

We developed a method for constructing personalized tumor progression models in terms of cancer hallmark acquisition and demonstrated the effectiveness of this model in terms of interpreting cancer evolutionary trees by means of an actual ccRCC dataset. In the example of ccRCC, identification of druggable target subclones that evolved after pathway alteration (HIF activation with a VHL mutation) is a clinically important problem. Our model has some implications. A cancer hallmark can help us to reduce complexity of cancer development and to characterize the phenotypes of subclones. Our method effectively incorporates cancer hallmarks into the current state-of-the-art tree reconstruction method of cancer subclonal evolution.

The three assumptions that we used to construct the model should be further examined in the context of cancer development. Our assumptions come from that an one time alteration of each pathway is enough for cancer establishment and development and a pathway alteration may trigger other biologically related pathway alterations. As a future challenge, we will examine our approach via exploring how the identified genomic level subclonal pathway alterations such as the common pathway alterations among the subclones, relate to downstream dysregulations, e.g., transcriptomic/proteomic level abnormality.

Acknowledgments. This work was supported by 16K16146, 16H01572, 15H05707, 18K18151, 18H04899 from The Japan Society for the Promotion of Science (http://www.jsps.go.jp/english/e-grants/grants01.html).

References

1. Nowell, P.C.: The clonal evolution of tumor cell populations. Science **194**, 23–28 (1976)
2. Gerlinger, M., Rowan, A.J., Horswell, S., Larkin, J., et al.: Intratumor heterogeneity and branched evolution revealed by multiregion sequencing. N Engl. J. Med. **366**, 883–92 (2012)
3. Popic, V., Salari, R., Hajirasouliha, I., Kashef-Haghighi, D., et al.: Fast and scalable inference of multi-sample cancer lineages. Genome Biol. **16**, 91 (2015)
4. Matsui, Y., Niida, A., Uchi, R., Mimori, K., Miyano, S., Shimamura, T.: PLoS Comput. Biol. **13**(5), e1005509 (2017)
5. Iorio, F., Alonso, L.G., Brammeld, J., Martincorena, I., et al.: Pathway-based dissection of the genomic heterogeneity of cancer hallmarks' acquisition with SLAPenrich. Sci. Rep. **8**, 6713 (2018)
6. Nik-Zainal, S., Alexandrov, L.B., Wedge, D.C., Van Loo, P., et al.: Mutational processes molding the genomes of 21 breast cancers. Cell **149**, 979–993 (2012)
7. Gerlinger, M., Horswell, S., Larkin, J., Rowan, A.J., et al.: Genomic architecture and evolution of clear cell renal cell carcinomas defined by multiregion sequencing. Nat. Genet. **46**, 225–233 (2014)
8. Iorio, F., Knijnenburg, T.A., Vis, D.J., Bignell, G.R., et al.: A landscape of pharmacogenomic interactions in cancer. Cell **166**, 740–754 (2016)

GIMLET: Identifying Biological Modulators in Context-Specific Gene Regulation Using Local Energy Statistics

Teppei Shimamura[1]([✉]), Yusuke Matsui[2], Taisuke Kajino[3], Satoshi Ito[4],
Takashi Takahashi[3], and Satoru Miyano[4]

[1] Division of Systems Biology, Nagoya University Graduate School of Medicine,
65 Tsurumai-cho, Showa-ku, Nagoya 466-8550, Japan
shimamura@med.nagoya-u.ac.jp
[2] Laboratory of Intelligence Healthcare, Nagoya University Graduate
School of Medicine, 1-1-20 Daiko-Minami, Higashi-ku, Nagoya 461-8673, Japan
[3] Division of Molecular Carcinogenesis, Nagoya University Graduate
School of Medicine, 65 Tsurumai-cho, Showa-ku, Nagoya 466-8550, Japan
[4] Human Genome Center, Institute of Medical Science, The University of Tokyo,
4-6-1 Shirokane-dai, Minato-ku, Tokyo 108-8639, Japan

Abstract. The regulation of transcription factor activity dynamically changes across cellular conditions and disease subtypes. The identification of biological modulators contributing to context-specific gene regulation is one of the challenging tasks in systems biology, which is necessary to understand and control cellular responses across different genetic backgrounds and environmental conditions. Previous approaches for identifying biological modulators from gene expression data were restricted to the capturing of a particular type of a three-way dependency among a regulator, its target gene, and a modulator; these methods cannot describe the complex regulation structure, such as when multiple regulators, their target genes, and modulators are functionally related. Here, we propose a statistical method for identifying biological modulators by capturing multivariate local dependencies, based on energy statistics, which is a class of statistics based on distances. Subsequently, our method assigns a measure of statistical significance to each candidate modulator through a permutation test. We compared our approach with that of a leading competitor for identifying modulators, and illustrated its performance through both simulations and real data analysis. Our method, entitled genome-wide identification of modulators using local energy statistical test (GIMLET), is implemented with R ($\geq 3.2.2$) and is available from github (https://github.com/tshimam/GIMLET).

Keywords: Gene regulation · Modulator detection ·
Energy statistics · Distance correlation · Statistical test

© Springer Nature Switzerland AG 2019
M. Bartoletti et al. (Eds.): CIBB 2017, LNBI 10834, pp. 124–137, 2019.
https://doi.org/10.1007/978-3-030-14160-8_13

1   Introduction

The regulation of gene expression is a process in which the expression of a particular gene can be either activated or repressed. Transcription factors (TFs) contribute greatly to the process of gene regulation by binding to a specific DNA sequence in the promoter regions of their target genes and controlling their transcription. The responsiveness of a target gene expression to a TF typically varies owing to genetic variation or a change in the cellular environment. This modulation in gene-specific responsiveness is often caused by a specific factor, called a modulator, at different levels, including the transcriptional, post-transcriptional and post-translational levels.

In the last decade, large international consortia, such as The Cancer Genome Atlas (TCGA) [1] and the International Cancer Genome Consortium (ICGC) [2], have generated large-scale gene expression profiles of different tumor types and catalogued their genetic alterations (recurrent mutations and copy number variations). Genome-wide association studies (GWAS) have also identified tens of thousands of human disease-associated variants and millions of single nucleotide polymorphisms [3]. However, it remains unknown if and in what way many genetic alterations and variants interact with physical and functional interactions within cellular networks.

The identification of genetic alterations and variations that function as biological modulators and contribute to gene expression control is one of the challenging tasks in systems biology. Recently, sophisticated algorithms have been developed for this task, which has successful applications in many areas [4–9]. For example, MINDy [4] formulated the problem of identifying modulators as a problem of testing whether the expressions of a univariate TF and its target gene, denoted by T and G, are independent each other, conditioned on the expression levels of an univariate modulator denoted by M in the framework of conditional mutual information. GEM [5] used a linear regression model with the effects of interaction between T and M to describe the relationships between T and G modulated by M. MIMOSA [6] considered a mixture model of T and G from two different fractions based on M. Note that these methods were designed to capture a particular type of three-way dependency, where T, G, and M are univariate random variables. Therefore, they cannot capture multivariate dependencies where sets of random variables are associated with each other. Currently, no systematic mathematical framework exists for identifying biological modulators of complex gene regulation, such as combinatorial regulation, where multiple TFs and modulators are functionally related.

In this study, we present a novel method, genome-wide identification of modulators using local energy statistical test (GIMLET), to overcome the challenges outlined above. GIMLET includes the following contributions.

1. GIMLET is mainly based on dependence coefficients from energy statistics for modeling the relationships between genes. These types of coefficients are a measure of the statistical dependence between two random variables or two random vectors of arbitrary, not necessarily equal dimension. This enables the

correlation of the expression of sets of any size for TFs, their target genes, and modulators.

2. We provide a new dependence coefficient, called local distance correlation, to compare the difference in distance correlation at low and high values of given modulators, allowing the identification of all types of local dependencies, such as nonmonotone and nonlinear relationships, between TFs and their target genes at the fixed point of modulators.

3. We develop a permutation-based approach to evaluate whether local distance correlation varies with modulators, which enables the discovery of modulators involved in complex regulatory relationships, including synergistic and cooperative regulation, from a statistical point of view.

We describe our proposed framework and algorithm in Sect. 2. We present the efficiency of GIMLET using synthetic and real data in Sects. 3 and 4.

2   GIMLET Methodology

2.1   Local Distance Correlation

The distance correlation [10,11] was introduced as a measurement of dependence between two random vectors $X \in \mathbb{R}^p$ and $Y \in \mathbb{R}^q$. It is based on the concept of distance covariance between X and Y which measures the distance between the joint characteristic function of (X, Y) and the product of the marginal characteristic functions. This method is extremely general in that it is applicable to random vectors of arbitrary and not necessarily equal dimension and only involves Euclidean pairwise distance. The remarkable properties of the distance correlation, denoted by $\mathcal{R}(X, Y)$, include $0 \leq \mathcal{R}(X, Y) \leq 1$ and $\mathcal{R}(X, Y) = 0$ if and only if X and Y are independent.

We introduce a local estimator of the distance correlation evaluated at another random vector $Z = z_\alpha \in \mathbb{R}^r$ as a local measurement of the dependence between X and Y conditioning on $Z = z_\alpha$ based on the observed data. We consider a collection $\{(x_k, y_k, z_k) : k = 1, \ldots, n\}$ of n i.i.d. observations for random vectors X, Y, and Z. Let us denote $w_{k\alpha} = K_h(z_k, z_\alpha)$ satisfying $\sum_{k=1}^n w_{k\alpha} = 1$ as the new weight function based on the distance between two sample vectors z_k and z_α where K_h is a specified kernel function with a bandwidth h.

Based on the definition of the Nadaraya-Watson estimator [12,13] as a weighted averaging method, we define a local estimator of distance covariance conditioning on $Z = z_\alpha$, using the weighted Euclidean distance as

$$\mathcal{V}_n^2(X, Y | Z = z_\alpha) = S_1(X, Y | Z = z_\alpha) + S_2(X, Y | Z = z_\alpha) - 2S_3(X, Y | Z = z_\alpha),$$

where

$$S_1(X, Y | Z = z_\alpha) = \sum_{k,l=1}^n w_{k\alpha} w_{l\alpha} |x_k - x_l| |y_k - y_l|,$$

$$S_2(X, Y | Z = z_\alpha) = \sum_{k,l=1}^{n} w_{k\alpha} w_{l\alpha} |\boldsymbol{x}_k - \boldsymbol{x}_l| \sum_{k,l=1}^{n} w_{k\alpha} w_{l\alpha} |\boldsymbol{y}_k - \boldsymbol{y}_l|,$$

$$S_3(X, Y | Z = z_\alpha) = \sum_{k=1}^{n} w_{k\alpha} \sum_{l,m=1}^{n} w_{l\alpha} w_{m\alpha} |\boldsymbol{x}_k - \boldsymbol{x}_l| |\boldsymbol{y}_k - \boldsymbol{y}_m|.$$

Each sample of the neighborhood in the α-th sample is weighted according to its weighted Euclidean distance from $Z = z_\alpha$. Points close to $Z = z_\alpha$ have a large weight, and points far from $Z = z_\alpha$ have a small weight. The kernel function K_h used in all of our examples is the Gaussian kernel function $K_h(z_k, z_\alpha) = \exp(-|z_k - z_\alpha|^2/h)$ where h is a bandwidth parameter that controls the smoothness of the fit. For a specific point $Z = z_\alpha$, the nearest-neighbor bandwidth h is determined such that the local neighborhood contains the $q = \lfloor n\delta \rfloor$ closest samples to the α-th sample in the Euclidean distance of Z, where $\delta \in (0, 1)$ is a tuning parameter that indicates the proportion of neighbors. Therefore, each local estimator is inferred with q observations that fall within the sphere $B_\delta(z_\alpha)$, centered at the α-th sample. We use a varying width parameter h that reduces the problem of data sparsity by increasing the radius in the regions with fewer observations.

The empirical local estimator of the distance correlation, called local distance correlation, $\mathcal{R}_n(X, Y | Z = z_\alpha)$ for given $Z = z_\alpha$ is then defined by the equation

$$\mathcal{R}_n(X, Y | Z = z_\alpha) = \frac{\mathcal{V}_n^2(X, Y | Z = z_\alpha)}{\sqrt{\mathcal{V}_n^2(X, X | Z = z_\alpha) \mathcal{V}_n^2(Y, Y | Z = z_\alpha)}} \tag{1}$$

if both $\mathcal{V}_n^2(X, X | Z = z_\alpha)$ and $\mathcal{V}_n^2(Y, Y | Z = z_\alpha)$ are strictly positive, and it is equal to zero otherwise.

2.2 Statistical Hypothesis Test for Identifying Modulators

In the statistical hypothesis testing for identifying modulators, the inference questions arise naturally such as whether the local dependence between X and Y are really varying with Z. This question can be formulated the hypothesis:

$$H_0 : \mathcal{R}_n(X, Y | Z) = c, \tag{2}$$

where $\mathcal{R}_n(X, Y | Z)$ is a function of Z and c is a constant.

For calculating the p-values of the local dependence between X and Y for each Z, we apply a permutation-based approach similar to the one used by [4]. Under the assumption that $\mathcal{R}_n(X, Y | Z)$ is a monotonic function of Z, we calculate the test statistic:

$$\Delta \mathcal{R}_n(X, Y | Z) = \left| \frac{1}{|\mathbb{U}_Z|} \sum_{k \in \mathbb{U}_Z} \mathcal{R}(X, Y | Z = z_k) - \frac{1}{|\mathbb{L}_Z|} \sum_{k \in \mathbb{L}_Z} \mathcal{R}(X, Y | Z = z_k) \right|, \tag{3}$$

where $\mathbb{U}_Z$ and $\mathbb{L}_Z$ are the index sets of the upper and lower points of Z, respectively. To assess the statistical significance of $\Delta \mathcal{R}_n(X, Y | Z)$, we generate a series of null hypotheses, and calculate the empirical p-value, using the following permutation procedures:

1. Permute the values of Z for all samples.
2. Re-calculate the test statistics using (3). Denote the null statistic of the l-th permutation by $\Delta \mathcal{R}_n^0(l)$.
3. Repeat steps 1–2 B times and calculate the empirical p-value for Z:

$$p_Z = \frac{1}{B} \sum_{l=1}^{B} I(\Delta \mathcal{R}_n(X, Y | Z) \le \Delta \mathcal{R}_n^0(l)), \tag{4}$$

where the indicator function $I(A)$ equals one when the condition A is true and it equals zero otherwise.

Note that this empirical method directly couples both the minimal obtainable p-value and the resolution of the p-value to the number of permutations B. Therefore, it requires a very large number of permutations to calculate the p-values when we want to accurately estimate small p-values. In order to compute more accurate p-values, we use a semi-parametric approach based on a tail approximation [14, 15]. The corrected empirical p-value $\tilde{p}_Z$, using the distribution tail approximation, is given by

$$\tilde{p}_Z = \begin{cases} p_Z & \text{if } \Delta \mathcal{R}_n(X, Y | Z) \le \Delta \tilde{\mathcal{R}}_n^0 \\ \exp\left[-\lambda(\Delta \mathcal{R}_n(X, Y | Z) - \Delta \tilde{\mathcal{R}}_n^0) \right] & \text{otherwise} \end{cases}, \tag{5}$$

where λ is a scale parameter, and $\Delta \tilde{\mathcal{R}}_n^0$ is a threshold that we set to the 99-th percentile of null distributions. The parameter λ is estimated by the null statistics satisfying the condition $\Delta \mathcal{R}_n^0 > \Delta \tilde{\mathcal{R}}_n^0$.

3   Synthetic Data Results

We generated synthetic data and evaluated the performance of our method in order to gain insight into the statistical power and type I error rate control in identifying modulators, based on the hypothesis $H_0 : \mathcal{R}_n(X, Y | Z) = c \leftrightarrow H_1 : \mathcal{R}_n(X, Y | Z) \ne c$.

A simulation study was conducted as follows. An i.i.d. sample of (X, Y, Z) was generated using the endogenous switching regression model under the following three settings:

$$\begin{aligned} M_1 : & \quad Y = \mu(X, Z) + \sigma(Z)\varepsilon, \\ M_2 : & \quad Y = \mu(X, Z) + \sigma(Z_1 Z_2)\varepsilon, \\ M_3 : & \quad Y = \mu(X_1 X_2, Z) + \sigma(Z)\varepsilon, \end{aligned}$$

with

$$\mu(X_1 X_2, Z) = \begin{cases} f_l(X_1 X_2) & \text{if } Z > \theta_1 \\ 0 & \text{otherwise} \end{cases}, \quad \text{and} \quad \sigma(Z) = \begin{cases} \gamma_1 & \text{if } Z > \theta_2 \\ \gamma_2 & \text{otherwise} \end{cases},$$

where $X, X_1, X_2, Z, Z_1, Z_2 \sim U[0, 1]$, $\varepsilon \sim N(0, 1)$, μ and σ are the conditional mean and variance of Y depending on Z, and f_l is a function that determines a functional relationship between X and Y.

For a function $f_l(X)$, we considered the following eight functional relationships:

$$
\begin{aligned}
&F_1\ (Line): &&f_1(X) = X - 1/2, \\
&F_2\ (Quadratic): &&f_2(X) = 4(X - 1/2)^2 - 1/2, \\
&F_3\ (Cubic): &&f_3(X) = 80(X - 1/3)^3 - 12(X - 1/3) - 7, \\
&F_4\ (Sinusoid,\ 2\ periods): &&f_4(X) = \sin(4\pi X), \\
&F_5\ (Sinusoid,\ 8\ periods): &&f_5(X) = \sin(16\pi X), \\
&F_6\ (x^{1/4}): &&f_6(X) = X^{1/4} - 1/2, \\
&F_7\ (Circle): &&f_7(X) = (2W - 1)\sqrt{1 - (2X - 1)^2}, \\
&F_8\ (Step): &&f_8(X) = I(X > 1/2) - 1/2,
\end{aligned}
\tag{6}
$$

where $W \sim Bern(0.5)$. These functions were originally used in [16] to assess the statistical power against independence.

We set θ_1 and θ_2 to be 0.25 and 0.75, respectively, and γ_1 and γ_2 as follows:

$$
\gamma_1 = \begin{cases} 1/6, \text{ if } f_l(X) = f_1(X) \text{ or } f_l(X) = f_3(X) \\ 1/2, \text{ otherwise} \end{cases},
$$

$$
\gamma_2 = \begin{cases} 1, \text{ if } f_l(X) = f_1(X) \text{ or } f_l(X) = f_3(X) \\ 3, \text{ otherwise} \end{cases}.
\tag{7}
$$

Scatter plots of the data obtained from these eight relationships are shown in Fig. 1.

The first setting, M_1, was designed to find modulators in the traditional framework for identifying modulators [4], where the expression value of a modulator $Z \in \mathbb{R}$ influences the dependence between the expression values of a TF $X \in \mathbb{R}$ and its target gene $Y \in \mathbb{R}$. The second and third settings, M_2 and M_3, were aimed at finding the modulators in the new conceptual framework investigated in this study: M_2 was intended for the combinatorial modulation, where the expressions of two modulators $Z = (Z_1, Z_2)' \in \mathbb{R}^2$ influence the dependency between a TF $X \in \mathbb{R}$ and its target gene $Y \in \mathbb{R}$. M_3 was intended for combinatorial regulation, where the expression of a modulator $Z \in \mathbb{R}$ influences the dependency between two TFs $X = (X_1, X_2)' \in \mathbb{R}^2$ and their target gene $Y \in \mathbb{R}$, and both X_1 and X_2 are required for Y.

The identification of modulators using our method (GIMLET) was assessed by comparing it with MINDy [4], one of the most widely used methods for this purpose. We note that MINDy cannot be directly applied to the identification of modulators under the settings M_2 and M_3, because MINDy is not designed for combinatorial modulation and regulation. In these simulations, all possible triplets were tested separately using MINDy, and the statistical significance was evaluated by using Fisher's method, which is widely used to combine p-values. A hypothesis testing problem for identifying modulators with varying sample sizes $(n = 100, 200, 500)$ was simulated with 1,000 datasets generated for each of the

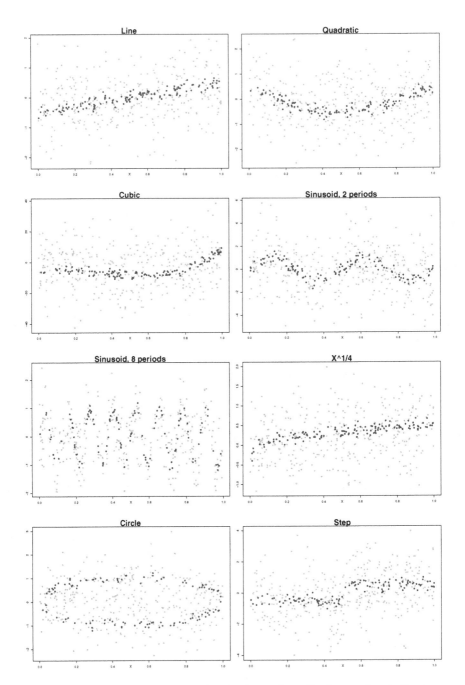

Fig. 1. Sample plots of the eight simulated relationships. Dark gray dots indicate samples with $Z > 0.75$, whereas light gray dots indicate samples with $Z \leq 0.75$.

above three settings. All tests were performed at the significance level $\alpha = 0.05$. The statistical power was estimated by the fraction of test statistics that were at least as large as the 95th percentile of the null distribution. The null distribution was calculated by $1,000$ permutations, as illustrated in Sect. 2. The type I error rate was estimated by calculating the power from data generated under the null hypothesis $H_0 : r(Z) = c$, which can be obtained by modifying the simulations where the random effect is set to be independent of Z. Theoretically, the type I error rate of the test should be equal to the significance level $\alpha = 0.05$.

Table 1. Statistical power of GIMLET and MINDy using synthetic data with different sample sizes ($n = 100, 200, 500$) for eight relationships (linear, quadratic, cubic, sine period $1/2$, sine period $1/8$, $x^{1/4}$, circle, and step), under three different settings (M_1, M_2, and M_3). The average of the p-values below the significant level $\alpha = 0.05$ were calculated through 1,000 simulations.

n	Relationship	Simulation model					
		M_1		M_2		M_3	
		GIMLET	MINDy	GIMLET	MINDy	GIMLET	MINDy
100	Line	**0.895**	0.576	0.621	0.090	0.652	0.141
	Quadratic	0.536	0.307	0.219	0.062	0.663	0.140
	Cubic	0.506	0.158	0.176	0.047	0.083	0.015
	Sine period $1/2$	0.345	0.271	0.164	0.065	0.141	0.041
	Sine period $1/8$	0.058	0.018	0.072	0.038	0.068	0.041
	$x^{1/4}$	0.750	0.314	0.241	0.055	0.708	0.200
	Circle	0.053	0.134	0.056	0.044	0.173	0.134
	Step	**0.880**	0.554	0.545	0.089	0.364	0.041
200	Line	**0.995**	0.939	**0.913**	0.121	**0.930**	0.315
	Quadratic	**0.861**	0.780	0.480	0.081	**0.935**	0.353
	Cubic	0.767	0.520	0.334	0.039	0.235	0.025
	Sine period $1/2$	0.679	0.463	0.336	0.061	0.297	0.040
	Sine period $1/8$	0.078	0.013	0.105	0.027	0.123	0.019
	$x^{1/4}$	**0.939**	0.693	0.474	0.043	**0.949**	0.405
	Circle	0.128	0.342	0.078	0.019	0.420	0.250
	Step	**0.994**	0.793	**0.866**	0.096	0.705	0.087
500	Line	**1.000**	1.000	**1.000**	0.208	**1.000**	0.761
	Quadratic	**0.995**	1.000	**0.885**	0.090	**1.000**	0.822
	Cubic	0.981	0.979	0.717	0.030	0.559	0.024
	Sine period $1/2$	**0.934**	0.997	0.759	0.087	0.709	0.067
	Sine period $1/8$	0.183	0.008	0.175	0.004	0.289	0.017
	$x^{1/4}$	**1.000**	0.996	**0.839**	0.035	**1.000**	0.850
	Circle	0.453	0.905	0.172	0.028	**0.916**	0.650
	Step	**1.000**	0.998	**0.997**	0.143	**0.965**	0.202

Table 1 shows the power calculated for eight different relationships with a varying sample sizes of 100, 200, and 500. Although both of the tested methods have low power in detecting modulators from a small sample size ($n = 100$), their power increases with the sample size. Note that GIMLET has higher power than MINDy in all relationships, except for the circle. Both GIMLET and MINDy have low chances of identifying modulators in the high-frequency sine relationship. GIMLET was shown to outperform MINDy, especially in the settings M_2 and M_3, because MINDy is not designed as a multivariate dependence measure for identifying modulators. Table 2 shows the type I error rates for the eight

Table 2. Type I error rate of GIMLET and MINDy, using synthetic data with different sample sizes ($n = 100, 200, 500$), for eight relationships (linear, quadratic, cubic, sine period 1/2, sine period 1/8, $x^{1/4}$, circle, and step), under three different settings (M_1, M_2, and M_3). The type I error rate of a test should be equal to the significance level $\alpha = 0.05$.

n	Relationship	Simulation model					
		M_1		M_2		M_3	
		GIMLET	MINDy	GIMLET	MINDy	GIMLET	MINDy
100	Line	**0.047**	**0.042**	**0.045**	0.071	**0.057**	0.061
	Quadratic	**0.041**	0.029	**0.052**	0.076	0.064	0.063
	Cubic	**0.050**	0.036	**0.051**	0.062	0.062	**0.051**
	Sine period 1/2	0.038	0.037	**0.058**	0.062	**0.053**	**0.053**
	Sine period 1/8	**0.045**	0.015	**0.041**	**0.051**	**0.057**	**0.040**
	$x^{1/4}$	**0.046**	0.023	**0.042**	**0.061**	**0.055**	**0.041**
	Circle	0.039	0.031	**0.056**	**0.055**	**0.049**	**0.056**
	Step	**0.053**	**0.049**	**0.048**	0.077	**0.054**	**0.056**
200	Line	**0.047**	0.035	**0.058**	0.066	**0.058**	0.034
	Quadratic	0.035	0.016	**0.052**	**0.059**	0.037	**0.044**
	Cubic	**0.045**	0.021	**0.048**	**0.046**	**0.043**	0.027
	Sine period 1/2	0.062	0.021	0.034	**0.047**	**0.043**	0.030
	Sine period 1/8	**0.049**	0.009	**0.045**	0.030	0.039	0.026
	$x^{1/4}$	**0.049**	0.030	**0.059**	**0.056**	**0.059**	**0.041**
	Circle	**0.059**	0.017	**0.048**	**0.056**	**0.046**	0.035
	Step	**0.046**	0.028	0.063	**0.058**	**0.055**	0.028
500	Line	**0.047**	0.021	**0.041**	0.030	**0.040**	0.012
	Quadratic	**0.051**	0.017	**0.053**	0.023	**0.046**	0.012
	Cubic	**0.046**	0.010	**0.048**	0.022	**0.047**	0.007
	Sine period 1/2	**0.060**	0.010	**0.053**	0.016	0.030	0.005
	Sine period 1/8	**0.053**	0.007	**0.053**	0.004	**0.045**	0.006
	$x^{1/4}$	**0.045**	0.018	**0.045**	0.024	**0.040**	0.013
	Circle	**0.056**	0.004	**0.049**	0.021	**0.053**	0.010
	Step	**0.046**	0.012	**0.044**	0.023	**0.047**	0.012

different relationships with varying sample sizes of 100, 200, and 500. The type I error rates are quite close to the chosen α level for all the tests, demonstrating that GIMLET shows better type I error rate control than MINDy, in this scenario.

4   Results with Real Data

We first sought to identify the genetic alterations that modulate the strength of the functional connection between HIF1A and the expression of its target genes in pan-kidney cohort in TCGA project [1]. The transcription factor HIF1A is a master transcriptional regulator of cellular and systemic homeostatic response to hypoxia. HIF1A activates the transcription of genes that are involved in crucial aspects of cancer biology, including angiogenesis, cell survival, glucose metabolism and invasion, and is implicated in the development of clear cell renal clear cell carcinoma (ccRCC). We examined mRNA expression profiles of 536 ccRCC and 357 non-ccRCC (papillary RCC and chromophobe RCC) patients, somatic mutation profiles of 436 ccRCC and 348 non-ccRCC patients, and copy number profiles of 528 ccRCC and 354 non-ccRCC patients, which can be downloaded from the Broad GDAC Firehose website [17]. We used 90 literature-validated target genes of HIF1A from the Ingenuity Knowledge Base [18] and calculated the factor scores for each patient by performing maximum-likelihood single factor analysis on the expression data matrix of these genes. In this example, we considered the factor score as the unobserved activity of HIF1A at the protein level and used it as Y. As candidates of Z, we first tested the somatic mutation of 85 genes, which were detected in more than 50 patients by genomic analyses of the pan-kidney cohort. We next considered the copy-number alterations of 41 chromosomal arms as candidates of Z. For this analysis, we expected to find an alteration of von Hippel-Lindau (VHL) tumor suppressor gene, which leads to overexpression of HIF1A and is a critical event in the pathogenesis of most ccRCC [19].

The modulator analysis of GIMLET yields five significantly associated gene mutations and genetic alterations modulating HIF1A activity with q-value < 0.10 (Table 3). Indeed, GIMLET identified VHL as the most significantly associated

Table 3. Five significantly associated gene mutations and genetic alterations modulating HIF1A activity.

Modulator	Type	q-value	ldcor (no mut/alt)	ldcor (mut/alt)
VHL	Mutation	0.001	0.24	0.49
3p	Deletion	0.001	0.23	0.44
20q	Amplification	0.001	0.42	0.20
20p	Amplification	0.002	0.42	0.20
PBRM1	Mutation	0.006	0.27	0.51

gene mutation. Although PBRM1, identified as the second-most significantly associated gene mutation, is not reported to directly modulate HIF1A activity, this result remains significant because almost all PBRM1 mutant cases also have dysregulation of the hypoxia signaling pathway [20] and it is likely that PBRM1 and VHL cooperate in kidney carcinogenesis, which leads to the overexpression of hypoxia-inducible HIF1A. The analysis also yields three regions significantly modulating HIF1A activity with a q-value < 0.10. Chromosome 3p deletions are observed in approximately 90% of ccRCC, which harbors VHL and tumor suppressor genes [21].

We next examined drug-treated gene expression profiles from Broad Institute The Library of Integrated Cellular Signatures (LINCS) Center for Transcriptomics [22]. We sought to use these data to identify drugs that inhibit the strength of the functional connection between FOXM1 and CENPF which are master regulators of prostate cancer malignancy [23] and the expression of their target genes. A total of perturbational gene expression profiles of 22,268 probes for 6,684 experiments treated with 271 compounds after 24 h under different doses (0.04, 0.12, 0.37, 1.11, 3.33, and 10 μm) in the two prostate cancer cell lines, PC3 and LNCaP, were downloaded from the LINCS L1000 dataset [22]. The expression values for each profile were normalized by robust z-scores relative to the control (plate population) and summarized using the median across replicates. If there are multiple probes that correspond to the same gene, the probe with the highest variance across all samples was selected as a single representative probe. Finally, the expression matrix data of 12,716 genes and 1,976 samples were used for further analysis. We used the expression of FOXM1 and CENPF as X and their unobserved activity as Y which was defined using maximum-likelihood single factor analysis on the expression data matrix for the 173 and 55 literature-validated targets of FOXM1 and CENPF, respectively, from the Ingenuity Knowledge Base [18]. The drug target genes for each compound under a given dose level were defined as differentially expressed genes, which were significantly lower in the drug-treated cell lines than in the vehicle-treated cell lines using a one-tailed t-test (p-value < 0.001). As candidates of Z, the drug-perturbational activity for each sample under each of 1,850 different pertubagens was then estimated using the enrichment scores (maxmean statistics) of these drug target gene sets for gene set analysis [24]. We applied GIMLET to identify functional pertubagens modulating FOXM1 and CENPF activity.

The analysis yields 13 pertubagens that significantly inhibit the regulation of FOXM1 and CENPF with a q-value $< 10^{-7}$ (Table 4). Indeed, these pertubagens support the inhibition of tumor progression in human prostate cancer by several resent studies. For example, Vorinostat known as suberanilohydroxamic acid is a member of a larger class of compounds that inhibit histone deacetylases (HDAC) [25]. A previous study has also shown that Vorinostat may inhibit tumor growth by both oral and parenteral administration in prostate cancer [26]. Withaferin A, a major bioactive component of the Indian herb Withania somnifera, induces cell death and inhibits tumor growth in human prostate cancer [27]. The activation of the PI3K-AKT-mTOR pathway is extremely common, if not universal,

Table 4. Thirteen significantly associated modulators (pertubagens) modulating FOXM1 and CENPF activity.

Modulator	Dose	Cell line	Target	$-\log_{10}(q\text{-value})$
Vorinostat	$10\,\mu$m	PC3	HDAC1	9.91
Withaferin A	$3.33\,\mu$m	PC3	MMP2	9.63
Dasatinib	$0.37\,\mu$m	PC3	ABL1	9.02
Dasatinib	$0.12\,\mu$m	PC3	ABL1	8.38
JW-7-24-1	$10\,\mu$m	PC3	LCK	8.38
OSI-027	$10\,\mu$m	PC3	mTOR	8.38
Radicicol	$10\,\mu$m	PC3	HSP90	8.38
PHA-793887	$3.33\,\mu$m	LNCaP	CDK2	8.30
WYE-125132	$10\,\mu$m	PC3	mTOR	8.07
GSK-1059615	$0.37\,\mu$m	PC3	PI3K	7.39
Sirolimus	$0.37\,\mu$m	LNCaP	mTOR	7.38
WYE-125132	$10\,\mu$m	PC3	mTOR	7.38
Celastrol	$1.11\,\mu$m	LNCaP	PSB5	7.20

in castrate-resistant prostate cancer [28]. Certain PI3K and mTOR inhibitors are currently under investigation in clinical trials for CRPC including the dual inhibitor NVP-BEZ235 [29] and the mTOR inhibitor RAD001 or everolimus [30,31].

The analyses with two examples thus show that GIMLET can identify genetic alterations and functional pertubagens modulating the relationship between a given set of regulators and the expression of their target genes in particular cancer subtypes.

5   Discussion

The identification of modulators is a challenging problem for researchers who study gene regulation. The paradigm introduced by [4] and the state-of-the-art classical methods for identifying modulators are quite useful because they allow us to identify content-specific modulators of a TF activity using gene expression data. However, these methods are restricted to the capturing of a particular type of dependency between univariate random variables, and it can be difficult to describe more complex multivariate dependency structures, when TFs and modulators are functionally related. We have developed a more general class of the identification of modulators, in the framework of energy statistics and a specific implementation, called GIMLET. An appealing property of the proposed method is that it can easily measure all types of dependencies, including nonmonotonic and nonlinear relationships, between random vectors in an arbitrary dimension. Our simulation results demonstrate that GIMLET outperforms

MINDy in terms of its statistical power and type I error rate. An analysis with a real example thus showed that GIMLET can identify genetic alterations and functional pertubagens modulating TF activities. We believe that the presented method may be useful for a range of biological applications, and it could represent a breakthrough in gene regulation research.

Acknowledgement. This work was supported by JSPS Grant-in-Aid for Challenging Exploratory Research (15K12139), JSPS Grant-in-Aid for Young Scientists A (15H05325), and JSPS Grant-in-Aid for Scientific Research on Innovative Areas (15H05912 and 18H04798). It was also supported in part by Ministry of Education, Culture, Sports, Science and Technology (MEXT) of Japan as a social and scientific priority issue (Integrated computational life science to support personalized and preventive medicine; hp170227, hp180198) to be tackled by using post-K computer. The super-computing resources were provided by Human Genome Center, University of Tokyo.

References

1. The Cancer Genome Atlas. https://cancergenome.nih.gov/
2. International Cancer Genome Consortium. http://icgc.org/
3. GWAS Catalog. https://www.ebi.ac.uk/gwas/
4. Wang, K., et al.: Genome-wide identification of post-translational modulators of transcription factor activity in human B cells. Nat. Biotechnol. **27**(9), 829–39 (2009)
5. Babur, Ö., et al.: Discovering modulators of gene expression. Nucl. Acids Res. **38**(17), 5648–56 (2010)
6. Hansen, M., et al.: Mimosa: mixture model of co-expression to detect modulators of regulatory interaction. Algorithms Mol. Biol. **5**, 4 (2010)
7. Alvarez, M.J., et al.: Functional characterization of somatic mutations in cancer using network-based inference of protein activity. Nat. Genet. **48**(8), 838–47 (2016)
8. Fazlollahi, M., et al.: Identifying genetic modulators of the connectivity between transcription factors and their transcriptional targets. Proc. Natl. Acad. Sci. U. S. A. **113**(13), E1835–43 (2016)
9. Hsiao, T.H., et al.: Differential network analysis reveals the genome-wide landscape of estrogen receptor modulation in hormonal cancers. Sci. Rep. **6**, 23035 (2016)
10. Székely, G.J., et al.: Measuring and testing dependence by correlation of distances. Ann. Statist. **35**(6), 2769–2794 (2007)
11. Székely, G.J., Rizzo, M.L.: Brownian distance covariance. Ann. Appl. Stat. **3**(4), 1236–1265 (2009)
12. Nadaraya, E.A.: On estimating regression. Theory Probab. Appl. **9**(1), 141–142 (1964)
13. Watson, G.S.: Smooth regression analysis. Indian J. Statist. Ser. A **26**(4), 359–372 (1964)
14. Knijnenburg, T.A., et al.: Fewer permutations, more accurate P-values. Bioinformatics **25**(12), i161–i168 (2009)
15. Matsui, M., et al.: D3M: detection of differential distributions of methylation patterns. Bioinformatics **32**(15), 2248–2255 (2015)
16. Simon, N., Tibshirani, R.: Comment on "detecting novel associations in large data sets". Science **334**(6062), 1518–1524 (2011)

17. The Broad GDAC Firehose. http://gdac.broadinstitute.org/
18. Ingenuity Knowledge Base. https://www.qiagenbioinformatics.com/products/ingenuity-pathway-analysis/
19. Maxwell, P.H., et al.: The tumour suppressor protein VHL targets hypoxia-inducible factors for oxygen-dependent proteolysis. Nature **399**(6733), 271–275 (1999)
20. Kapur, P., et al.: Effects on survival of BAP1 and PBRM1 mutations in sporadic clear-cell renal-cell carcinoma: a retrospective analysis with independent validation. Lancet Oncol. **14**(2), 159–167 (2013)
21. Bregarolas, J.: Molecular genetics of clear-cell renal cell carcinoma. J. Clin. Oncol. **32**(18), 1968–1976 (2014)
22. The Library of Integrated Cellular Signatures. http://www.lincsproject.org/
23. Lokody, I.: Signalling: FOXM1 and CENPF: co-pilots driving prostate cancer. Nat. Rev. Cancer **14**(7), 450–451 (2014)
24. Efron, B., Tibshirani, R.: On testing the significance of sets of genes. Ann. Appl. Stat. **1**(1), 107–129 (2007)
25. Wikipedia. https://en.wikipedia.org/wiki/Vorinostat
26. Bulter, L.M., et al.: Suberoylanilide hydroxamic acid, an inhibitor of histone deacetylase, suppresses the growth of prostate cancer cells in vitro and in vivo. Cancer Res. **60**, 5165–5170 (2000)
27. Yang, H., et al.: The tumor proteasome is a primary target for the natural anti-cancer compound Withaferin A isolated for "Indian winter cherry". Mol. Pharmacol. **71**, 426–437 (2007)
28. Lian, F., et al.: The biology of castration-resistant prostate cancer. Curr. Probl. Cancer **39**(1), 17–28 (2015)
29. Hong, S.W., et al.: NVP-BEZ235, a dual PI3K/mTOR inhibitor, induces cell death through alternate routes in prostate cancer cells depending on the PTEN genotype. Apoptosis **19**(5), 895–904 (2014)
30. Nakabayashi, M., et al.: Phase II trial of RAD001 and bicalutamide for castration-resistant prostate cancer. BJU Int. **110**(11), 1729–1735 (2012)
31. Templeton, A.J., et al.: Phase 2 trial of single-agent everolimus in chemotherapy-naive patients with castration-resistant prostate cancer (SAKK 08/08). Eur. Urol. **64**(1), 150–158 (2013)

Structural Features of a DPPG Liposome Layer Adsorbed on a Rough Surface

Maria Raposo[1](✉) ⓘ, Andreia A. Duarte[1], Paulo J. Gomes[1],
Paulo A. Ribeiro[1] ⓘ, Marli L. Moraes[2] ⓘ, and Roland Steitz[3]

[1] CEFITEC, Departamento de Física, Faculdade de Ciências e Tecnologia, FCT,
Universidade Nova de Lisboa, 2829-516 Caparica, Portugal
mfr@fct.unl.pt
[2] Universidade Federal de São Paulo, Campus São José dos Campos,
Rua Talim, 330, Jardim Aeroporto, São José dos Campos, SP 12231280, Brazil
[3] Helmholtz-Zentrum Berlin für Materialien und Energie,
Hahn-Meitner-Platz 1, 14109 Berlin, Germany

Abstract. The development of drug delivery systems, sensors and other devices based on liposomes (small unilamellar lipid vesicles, SUVs) requires the adsorption of intact lipid structures onto solid surfaces in the first place. In this work, we report on the in situ investigation of the adsorption of liposomes of 1,2-dipalmitoyl-*sn*-glycero-3-[phospho-*rac*-(1-glycerol)] (sodium salt) (DPPG) onto a rough surface by neutron reflectivity. Rough surfaces are achieved by preparing polyelectrolyte layer-by-layer films, which act as soft polymer cushions. Neutron reflectivity measurements performed at the solid/D_2O interface allow for the determination of the thickness of the adsorbed structures. The conducted investigation proofs that the liposomes dispersed in the liquid phase are generally adsorbed intact onto the cushion surface, confirming that the roughness of the latter is a variable to be taken into account if one intends to adsorb intact lipid structures. Liposome flattening is observed and justified by the attractive electrostatic interactions occurring between the negatively charged lipid liposomes and the outermost, positively charged polyelectrolyte layer of the cushion. The conducted measurements further demonstrate that the adsorbed liposomes are stable for several hours. These findings are fundamental for the development of devices based on immobilized but intact SUVs on sensor surfaces.

Keywords: Liposome · Surface · Adsorption · Neutron reflectivity · Roughness

1 Introduction

Artificial cells or protocells based on natural biomolecules and synthetic compounds can be assembled in a bottom-up approach by preparing lipid vesicles or liposomes carrying biological molecules inside. Although the use of such lipid containers in the dedicated engineering of cellular machinery is still a rather experimental approach, the preparation of liposomes itself is already well established [1–4]. Prototypical examples based on implemented liposome systems are biosensors, drug delivery systems and

M. Bartoletti et al. (Eds.): CIBB 2017, LNBI 10834, pp. 138–144, 2019.
https://doi.org/10.1007/978-3-030-14160-8_14

encapsulated molecules systems for irradiation studies with high energy photons (or particle beams) in a cellular environment. Most of these applications require that the liposomes/vesicles to be adsorbed stay intact on their solid supports. First studies of liposomes adsorption demonstrated that the liposomes opened upon contact with the solid surface and transformed into adsorbed lipid bilayers [5 and references therein], [6]. Recently, a study conducted by Duarte and co-workers revealed that the morphology of the surface might influence liposome integrity and also determine adsorption kinetics [7]. These suggestions are supported by quartz crystal microbalance (QCM) and atomic force microscopy (AFM) measurements [7].

In the work reported here, we used neutron reflectivity for the in situ characterization of the interfacial structure of a polymeric cushion prepared by the layer-by-layer (LbL) technique [6 and references therein] before and after adsorption of liposomes of 1,2-dipalmitoyl-*sn*-glycero-3-[phospho-*rac*-(1-glycerol)] (sodium salt) (DPPG) against the liquid phase on the molecular scale. Special care was taken in generating a rough surface of the polymer cushion. As a remark, the lipid utilized is one of the most investigated to model membranes and presents a negative charge at pH ∼ 7 allowing to be adsorbed onto positively charged surfaces. In the present work, DPPG was used for comparison with the findings from a very similar system investigated in an earlier study utilizing quartz crystal microbalance, atomic force microscopy, and X-ray photoelectron spectroscopy [6].

2   Materials and Methods

Synthetic   1,2-dipalmitoyl-*sn*-glycero-3-[phospho-*rac*-(1-glycerol)]   (sodium   salt) (DPPG) lipid, with molecular weight of 744.96 $gmol^{-1}$, was purchased from Avanti Polar Lipids. Small unilamellar vesicles (SUVs) were prepared by dissolving 5 mM DPPG in methanol:chloroform (2:8). After solvent evaporation using a gentle nitrogen stream, the lipid film was hydrated overnight in pure water supplied by a Milli-Q purification system (resistivity 18.2 MΩcm and pH ∼ 5.7). The solution was vortexed intermittently leading to multilamellar vesicles (MLVs). The SUVs (liposomes) were then obtained by extruding this solution in a mini-extruder from Avanti Polar Lipids in a polycarbonate membrane with 0.1 μm pores. The number of passages through the membrane were eleven. Figure 1 shows the chemical structure of DPPG molecule as well as those of the polyelectrolytes used in the preparation of the soft polymer cushion.

The liposomes were adsorbed onto the polymeric cushion which was obtained by the LbL technique and which was pre-deposited onto a silicon wafer. The polymeric cushion resulted from the alternated adsorption of poly(ethylene imine) (PEI), poly (allylamine hydrochloride) (PAH) and poly(styrene sulfonate) (PSS) polyelectrolytes from respective aqueous solutions with monomeric concentrations of 10^{-2} M. The cushion layer sequence was PEI/(PSS/PAH)$_5$. The adsorption period was 30 min for the PEI layer and 20 min for the PAH and PSS layers. Between each adsorption step wafer + LbL film (cushion) were rinsed with ultrapure water. The silicon wafer of $80 \times 50 \times 10$ mm^3 was purchased from Holm Siliciumbearbeitung, Tann, Germany, and cleaned with "piranha" solution containing hydrogen peroxide and sulfuric acid

(1:1). It was kept in the piranha solution for 30 min prior to polymer adsorption, after which it was exhaustively washed with ultrapure water. To avoid contamination, the wafer was stored in ultrapure water until the sample preparation.

The cushion was characterized by neutron reflectivity measurements at the solid/liquid interface before and after lipid adsorption with a neutron beam of wavelength 4.66 Å, a graphite monochromator and a 3He detector, on the V6 reflectometer facility at the Berlin Neutron Scattering Center (BENSC), Helmholtz Zentrum Berlin für Materialien und Energie (former Hahn-Meitner-Institut). The measurements consisted in obtaining the reflectivity patterns of the thin films at the solid/D_2O interface. The used experimental setup is similar to that described by Howse et al. [8] and Steitz et al. [9]. The data were fitted by applying the optical matrix method [10] with the Parratt32 fitting program [10, 11] which allowed for determining layer structure and thickness.

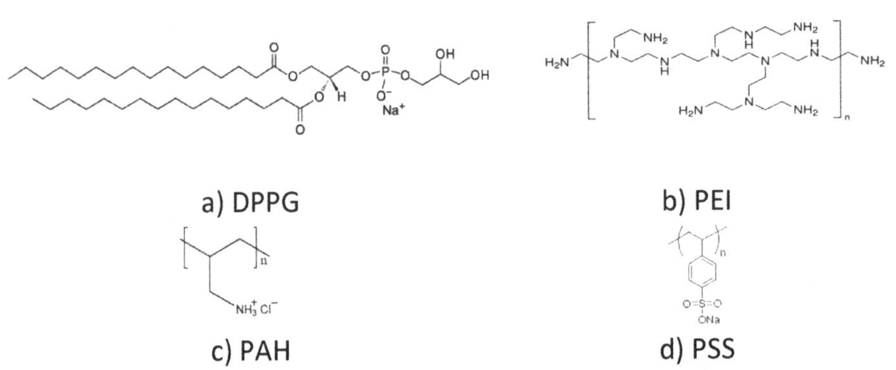

Fig. 1. Chemical structure of (a) 1,2-dipalmitoyl-sn-glycero-3-[phospho-rac-(1-glycerol)] (sodium salt) (DPPG); (b) Poly(ethyleneimine) (PEI); (c) poly(allylamine hydrochloride) and (d) Poly (styrene sulfonate) (PSS).

3 Results

Figure 2 shows the neutron reflectivity curves at the solid/D_2O interface of the PEI/(PSS/PAH)$_5$ LbL film (cushion) without and with liposomes adsorbed to it. The method of characterization of these curves and the achievement of respective parameters is reported in [12]. Several attempts have been carried out to fit the experimental data. Best match was achieved with a 4 box model representing the polymer cushion with scattering length density (SLD) and roughness increasing towards the liquid phase. The total thickness of 348 Å obtained was slightly above the values measured for similar, but flat cushions [9, 12]. The observed increase in thickness is explained by a more imperfect coating with continuously increased number of internal voids and roughness towards the liquid phase. Successful adsorption of DPPG is proven by the respective shift of the so-called Kiessig fringes in the reflectivity pattern to the left (lower Q) and respective enhancement in amplitude (Fig. 2). Interestingly, all attempts to fit simple models based on lipid bilayer structures adsorbed to the polymer cushion failed.

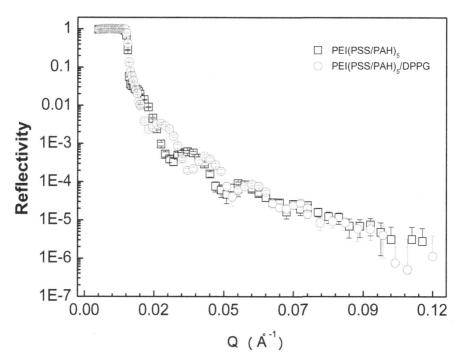

Fig. 2. A Neutron reflectivity curves as obtained from the PEI(PSS/PAH)$_5$ (black) and PEI (PSS/PAH)$_5$/DPPG films (green). (Color figure online)

Conducting a simple analysis by multiplying the reflectivity data with Q^4 and plotting the minima positions Q_n versus their order n, the extracted thickness from the coatings (linear fits with slope $d/(2\pi)$) was 354 ± 28 Å for the cushion and 428 ± 8 Å for the cushion + DPPG. That is to say that adsorption of DPPG vesicles resulted in a net increase in thickness of 74 ± 36 Å. The thickness of a DPPG bilayer in dry state is 43 ± 2 Å. This value corresponds to the distance between the phosphorus peaks across the bilayer and was calculated using simulated electron density profiles [13, 14]. The measured value in the present work is smaller but is approaching the thickness of two individual phospholipid bilayers. In fact, the thickness of a unilamellar lipid vesicle adsorbed in pancake fashion must be at minimum 86 Å, with very little free water between the lipid leaflets and with additional swelling in liquid environment not taken into account. The neutron reflectivity measurements, however, were carried out at solid/liquid interface with deuterated water further hydrating the lipid bilayers. As still the measured thickness of the adsorbed liposomes is smaller than 86 Å one has to conclude that severe structural reorganizations of the vesicles in adsorbed state must have occurred, for instance strong tilting of the aliphatic lipid chains against the surface normal.

Duarte et al. [7] have shown that the fraction of open liposomes decreases with cushion roughness. At a surface roughness of the cushion of about 40 Å, the fraction of open DPPG liposomes was 0.5. At a surface roughness of 60 Å the fraction of open DPPG liposomes was zero, i.e. DPPG liposomes adsorbed exclusively in intact state [7]. In the present work, the cushion roughness as measured by neutron reflectivity is 58 ± 9 Å. Therefore, in accordance with ref. [7], it is expected that liposomes are adsorbed intact. Although the obtained values do not allow excluding that a small fraction of adsorbed liposomes is open, most liposomes must have adsorbed intact but flattened on the cushion surface. The liposome flattening is justified by the attractive electrostatic interactions occurring between the negatively charged lipids and outer-most, positively charged polyelectrolyte layer of the cushion. Figure 3 shows a schematic representation of liposomes on the cushion surface. The results obtained in this work are consistent with those obtained by QCM and AFM techniques and allow to conclude that roughness of the receiving surface is a variable determining the structure and in consequence the functionality of the immobilized lipid aggregates. Finally, as the reflectivity measurements take several hours, the obtained results also reveal that the adsorbed liposomes are stable at least during the time period of measurements.

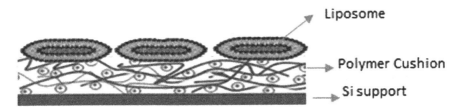

Fig. 3. Oversimplified schematic representation of the PEI(PSS/PAH)$_5$/DPPG films (not to scale).

4   Conclusions

This work shows that surface roughness is a key parameter in the achievement of intact liposomes immobilized on solid surfaces. The adsorbed liposomes are found stable for several hours. The adsorbed liposomes take on a flattened conformation (pan cake type). Attractive electrostatic interactions occurring between the lipids and the outer-most layer of the cushion are held responsible for the pan cake conformation of the immobilized DPPG aggregates. The reported findings are of extreme importance for the development of drug delivery systems, sensors and other devices, based on liposomes as in particular intact adsorbed liposomes can keep encapsulated molecules hydrated and in a medium close to the cellular environment.

Finally, as concluding remarks, this work is an example of nano-bio-technology associated to Bioinformatics and Biostatistics for the following reasons: (1) the obtained structures lies in the nm range; (2) liposome thin films can be employed as biosensors having the advantage of liposomes surface properties (recognition capacity, selectivity, etc.) can be easily tuned by varying the lipid composition; (3) by applying the concept of electronic tongue [15] to the liposome biosensors, the methods of data treatment are the ones used in the Bioinformatics and Biostatistics scientific area; (4) adsorption of liposome with encapsulated molecules on scaffolds can allow controlled delivery of drugs during, for example, tissue growth; and (5) the adsorption of intact liposome is a contribution for development of synthetic biology with the next step being the study of the conditions leading to adsorption of intact exosomes on surfaces.

Acknowledgments. The authors acknowledge the V6 reflectometer facility at the Berlin Neutron Scattering Center (BENSC), Helmholtz Zentrum Berlin für Materialien und Energie (former Hahn-Meitner-Institut), Berlin, Germany and for their support through the Access to Research Infrastructure action. The authors acknowledge financial support from FEDER, through Programa Operacional Factores de Competitividade − COMPETE and Fundação para a Ciência e a Tecnologia − FCT, by the project PTDC/FIS-NAN/0909/2014 and the Portuguese research Grant No. PEst-OE/FIS/UI0068/2011 and UID/FIS/00068/2013 through FCT-MEC (Portugal) and by FAPESP and CNPq (Brazil).

References

1. Luisi, P.L., Ferri, F., Stano, P.: Approaches to semi-synthetic minimal cells: a review. Naturwissenschaften **93**(1), 1–13 (2006)
2. Lukyanov, A.N., Torchilin, V.P.: Micelles from lipid derivatives of water-soluble polymers as delivery systems for poorly soluble drugs. Adv. Drug Deliv. Rev. **56**(9), 1273–1289 (2004)
3. Akbarzadeh, A., et al.: Liposome: classification, preparation, and application. Nanoscale Res. Lett. **8**, 102 (2013). https://doi.org/10.1186/1556-276x-8-102
4. Jesorka, A., Orwar, O.: Liposomes: technologies and analytical applications. Ann. Rev. Anal. Chem. **1**, 801–832 (2008)
5. Pires, F., Duarte, A., Ferreira, Q., Magalhães-Mota, G., Ribeiro, P.A., Raposo, M.: Imaging of liposomal drug delivery systems by atomic force microscopy. In: Méndez-Vilas, A. (ed.) Microscopy and Imaging Science: Practical Approaches to Applied Research and Education. pp. 183–194. Formatex Research Center (2017). http://www.microscopy7.org/
6. Haas, H., Raposo, M., Ribeiro, P., Steitz, R., Cavatorta, P., Riccio, P.: Myelin model membranes on a soft polymer cushion. Biophys. J. (Annual Meeting Abstracts), **80**(1), Part 2, 23A–24A (2001)
7. Duarte, A.A., et al.: DPPG liposomes adsorbed on polymer cushions: effect of roughness on amount, surface composition and topography. J. Phys. Chem. B. **119**(27), 8544–8552 (2015)
8. Howse, J.R., Manzanares-Papayanopoulos, E., McLure, I.A., Bowers, J., Steitz, R., Findenegg, G.H.: Adsorption from alkane+perfluoroalkane mixtures at fluorophobic and fluorophilic surfaces. II. Crossover from critical adsorption to complete wetting. J. Chem. Phys. **116**, 7177 (2002)
9. Steitz, R., Leiner, V., Siebrecht, R., Klitzing, R.: Influence of the ionic strength on the structure of polyelectrolyte films at the solid-liquid interface. Colloids Surf. A Physicochem. Eng. Aspects **163**, 63–70 (2000)
10. Parrat, L.G.: Surface studies of solids by total reflection of X-rays. Phys. Rev. **95**, 359 (1954)
11. Russel, T.P.: X-ray and neutron reflectivity for the investigation of polymers. Mater. Sci. Rep. **5**, 171–271 (1990)

12. Ribeiro, P.A., Steitz, R., Lopis, I.E., Haas, H., Oliveira Jr., O.N., Raposo, M.: Thermal stability of poly(o-methoxyaniline) layer-by-layer films investigated by neutron reflectivity and UV-VIS spectroscopy. J. Nanosci. Nanotechnol. **6**, 1396–1404 (2006)
13. Zaraiskaya, T., Jeffrey, K.R.: Molecular dynamics simulations and 2H NMR study of the GalCer/DPPG lipid bilayer. Biophys. J. **88**, 4017–4031 (2005)
14. Pimthon, J., Willumeit, R., Lendlein, A., Hofmann, D.: All-atom molecular dynamics simulation studies of fully hydrated gel phase DPPG and DPPE bilayers. J. Mol. Struct. **921**, 38–50 (2009)
15. Magro, C., Mateus, E., Raposo, M., Ribeiro, A.: Overview of electronic tongue sensing in environmental aqueous matrices: potential for monitoring emerging organic contaminants. Environ. Rev. (2018). http://dx.doi.org/10.1139/er-2018-0019

Chemical Exchanges and Actuation in Liposome-Based Synthetic Cells: Interaction with Biological Cells

Giordano Rampioni[1], Francesca D'Angelo[1], Alessandro Zennaro[1],
Livia Leoni[1], and Pasquale Stano[2]($\boxtimes$)

[1] Sciences Department, Roma Tre University,
Viale G. Marconi 446, 00146 Rome, Italy
[2] Department of Biological and Environmental Sciences and Technologies (DiSTeBA),
University of Salento, Ecotekne, 73100 Lecce, Italy
`pasquale.stano@unisalento.it`

Abstract. The development of new synthetic biology frontiers has led to scenarios where the embodied information-processing capability of biological organisms are implanted, in minimalistic version, in liposome-based synthetic cells. These are cell-like systems of minimal complexity resembling biological cells. Although not yet alive, synthetic cells are useful for generating basic biological understanding, and can become interesting biotechnological tools. In 2012 we devised a research program aimed at the design and construction of synthetic cells capable of exchanging chemical signals with biological micro-organisms (in particular bacteria). Here we review the fundamental steps leading to this innovative research field and comment on the most relevant experimental results obtained by us and others.

Keywords: Bio-chem ITs · Lipid vesicles ·
Molecular communications · Synthetic Biology · Synthetic Cells

1 Lipid Vesicles in Origin-of-Life Research and in Synthetic Biology

The goal of synthetic biology (SB) is the design and the construction of biological parts, devices and programmable systems to perform useful functions. In the majority of cases these novel synthetic parts are implemented in living cells, taking advantage of their native cellular 'chassis' (i.e., the genome, the native set of transcription factors, the pre-existing metabolic routes, the protein-functionalized membrane, and so on). There is however another approach which focuses on the total assembly of synthetic cells (SCs) starting from separated biological parts [29,38,40,46]. This second approach, often labeled as *bottom-up*, leads to very simple cell-like structures (yet not living), by exploiting the convergence of three main technologies: (a) cell-free systems, in particular

© Springer Nature Switzerland AG 2019
M. Bartoletti et al. (Eds.): CIBB 2017, LNBI 10834, pp. 145–158, 2019.
https://doi.org/10.1007/978-3-030-14160-8_15

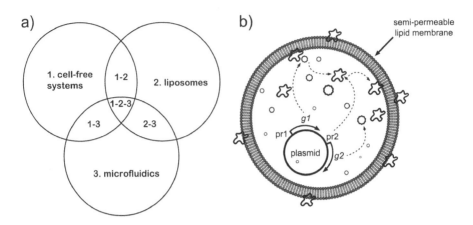

Fig. 1. Semi-synthetic (minimal) cells. (**a**) Three of the technological pillars for the bottom-up construction of liposome-based SCs, namely: cell-free systems, liposome technology, microfluidics. A brief referenced description of these three technologies and of their intersections is reported in Appendix A. (**b**) Schematic drawing of semi-synthetic minimal cells, as derived from the inclusion of cell-free systems (water-soluble proteins, membrane proteins, the TX-TL system, DNA in form of a plasmid, and low molecular-weight compounds). The shown structure can be obtained experimentally by combining liposome technology and cell-free system (or by the use of microfluidic devices). A great impulse to the construction of cell-like systems as depicted in this figure has resulted by the introduction of a special giant vesicle formation method, namely, the droplet-transfer method [37] which allow the easy encapsulation of complex biochemical machineries inside vesicles that have size in the 5–20 μm range (typically), thus easily observable under the optical microscope.

transcription-translation (TX-TL), (b) lipid vesicles or similar compartments, (c) microfluidics (Fig. 1a).

The first experiments for constructing simplified cells in the laboratory started within the community of origin-of-life researcher, as a way to study primitive cells, inspired by the Haldane-Oparin "principle of continuity" between inanimate and animate matter [35]. According to this principle, living systems are just a type of molecular assemblies; despite their complexity, they are not qualitatively different than any other chemical system. This premise implies that the scientific progresses will probably allow the construction of a living system in the laboratory, neglecting, then, any vitalistic approach. One of the theoretical frameworks that can guide the construction of minimal living systems is the autopoietic theory [28], a system view of life that specifically describes living systems as those systems whose built-in and observer-independent 'purpose' is just their own self-maintenance.

Among the various possible bottom-up SCs designs, the so-called semi-synthetic approach [23] appears to be one of the most promising in terms of feasibility, versatility, modularity, robustness, and possibility of interfacing with biological systems. Semi-synthetic minimal cells can be defined as those synthetic cell-like systems based on the encapsulation of the minimal number of biomolecules (nucleic acids, proteins, etc.) inside lipid vesicles (liposomes), see Fig. 1b, and that display minimal living-like properties. For example, these properties can be: autopoietic self-maintenance, self-reproduction, movement, signal recognition, and so on. As specified, current approaches have not lead to living SCs yet. Nevertheless these sort of "intermediate" structures – between life and non-life – are useful in many theoretical and practical respects.

We have recently proposed that semi-synthetic SCs of minimal complexity can play a major role in the nascent field of Molecular Communications [32] and more in general of bio-chemical Information and Communication Technologies (bio-chem ICTs) [43]. Bio-chem-ICTs can be defined as those technologies deriving from the integration of research areas such systems- and synthetic biology, micro-electromechanical systems, chemical information processing and nanobiotechnology. Their applications range from the creation of life-inspired technologies to smart drug delivery, from adaptive artificial systems to novel information-energy-matter processing [3]. In a narrower context, and inspired by the natural signal processing ability of biological cells, we foresee that SCs can manipulate chemical information in a programmable way by reconstructing a minimal set of molecular sensors, actuators, controllers inside liposomes. After decades of electromechanical-based systems, cybernetics can thus re-start in the biomolecular realm [41].

2 The Technology of SCs

Figure 1a and Appendix A evidence which are, in the authors opinion, the three most relevant technologies currently used to build SCs, in particular liposome-based SCs that operate by gene expression.

Lipid vesicles (liposomes) have played and still play a central role in this enterprise. Gradually, the research moved from liposomes encapsulating simple chemicals to liposomes encapsulating biochemical reactions, and reached the contemporary approach that is strongly anchored to intra-liposome TX-TL reactions. In particular, by means of TX-TL it is possible to produce proteins in order to functionalize SCs. The *in situ* produced proteins are 'actuators', which can perform specific operations in the SCs such as catalyzing reactions, binding small molecules, exerting control over the permeation of other molecules across the lipid membrane, regulating TX, acting as cytoskeleton elements, and so on.

When SCs are prepared by traditional vesicle formation methods, a large between-vesicle variability is observed, especially with respect to solute encapsulation [2]. This translates into a large heterogeneity in SC function (rate of protein production, for example). Such phenomenon is due to extrinsic stochastic effects due to solute partition in the very moment of vesicle formation. Microcompartments, in fact, contain – on average – a small number of solutes and therefore the internal solute concentration can vary either expectedly [42] or unexpectedly [24].

When this sort of spontaneously emerged 'vesicle diversity' is not desirable, microfluidic technology becomes one of the best tools for SC technology. To date, most of published papers report on SCs constructed by traditional batch methods (sector *1–2* of Fig. 1a), but an ever-increasing number of recent reports focuses on the employment of microfluidic chips to build solute-filled vesicles (generally giant vesicles), realizing, *de facto*, the next standard for SCs technology. These advancements will lead, once properly developed, to a sort of assembly line machine for producing SCs with low between-vesicle variability.

A fourth subject, not explicitly included in Fig. 1a, is mathematical modeling. According to SB paradigm, the construction of synthetic parts, device and systems is always strongly integrated with a design phase, which requires knowledge on chemistry, biochemistry and bioinformatics. From this consideration, it emerges a pressing requirement, i.e., understanding biological cellular processes at a quantitative level. Values of binding constants, kinetic constants, and connectivity between the network components are essential ingredients of accurate models, as well as the inclusion of stochastics processes (intrinsic and extrinsic). The network dynamics is another related topic that can be grafted into SC research for a better SC understanding and design (e.g., synchronization phenomena, essential in both self-reproduction and self-maintenance [12]). Among the several goals of modeling, here we recall that novel approaches based on information and communication theories should be properly developed in molecular communications, taking into account that chemical signals are often based on passive stochastic diffusion [31].

Bottom-up synthetic biology represents a very modern, challenging and promising arena for developing novel systems of increasing complexity, from SCs to synthetic cellular communities, encompassing basic and applied science. There are studies where SCs are used to understand biochemical phenomena, or to provide models of intermediate complexity (before more complex SCs are created), or to play a role for biotechnological applications. Research in this field is moving from very simple SCs to systems capable of performing more complicated tasks, which often result from the integration of previously achieved performances. In other words, the contemporary studies aim at demonstrating that SCs can go beyond the proof-of-principle stage that characterized the "pioneer" phase [45].

Recent examples demonstrate the progresses of SC research. It has been reported that SCs can synthesize a therapeutic protein inside the tumor [19], or that a modular design can bring to artificial platelets [25]. Enzymatic fermentation-like reactions [13], as well as the *de novo* synthesis of cell division proteins have been achieved [14]. DNA has been replicated in SCs [34]. Thanks to microfluidics, it has been possible to achieve a sequential bottom-up assembly of SCs [50].

A particularly elaborated and elegant system, published in 2008, has been reported by Tetsuya Yomo and collaborators, who constructed a cell-like system capable of synthesizing the $Q\beta$-replicase enzyme starting from a RNA sequence, by a TL reaction [18]. The same RNA sequence, however, is also a substrate of $Q\beta$-replicase, that builds the complementary strand of it. These reactions originate a recursive process, whereby a RNA molecule encodes for the enzyme, whose function is the RNA replication. This work can be seen as an attempt of constructing cell-like systems where nucleic acids and proteins reciprocally induce their own formation.

3   Chemical Signaling and Actuation Control of SCs

The Vision

The progress and the success of SC technologies has motivated several scholars to inquire whether and at what extent it is possible to engineer SCs so to allow chemical signaling and actuation (i.e., execute a predetermined operation in response to a well defined stimulus).

The interest in such a kind of SCs functionalization is manifold, but it mainly focuses on the construction of programmable SCs that "do something" only when it is required. In other terms, we are starting to conceive very simple soft-wet-microrobots that share with biological systems their constituents, the structure, and – importantly – the mechanism of the operations.

The appeal of such a vision was recognized by Arturo Rosenblueth, Norbert Wiener and Julian Bigelow, the pioneers of cybernetics, who in their 1943 foundation paper wrote [41]:

> If an engineer were to design a robot, roughly similar in behavior to an animal organism, he would not attempt at present to make it out of proteins and other colloids. He would probably build it out of metallic parts, some dielectrics and many vacuum tubes. The movements of the robot could readily be much faster and more powerful than those of the original organism. Learning and memory, however, would be quite rudimentary. In future years, as the knowledge of colloids and proteins increases, future engineers may attempt the design of robots not only with a behavior, but also with a structure similar to that of a mammal. The ultimate model of a cat is of course another cat, whether it be born of still another cat or synthesized in a laboratory.

Now, we "future engineers" have the chance of designing and constructing robots – at the supramolecular cell-like scale – thanks to our increased knowledge of "proteins and colloids", accumulated after decades of research on biochemistry and molecular biology.

It is evident that a reachable goal consists in SCs capable of sending and receiving chemical messages, and possibly activating internal genetic circuits that ultimately lead to a specific response – usually the synthesis of a protein.

This topic emerged, independently, from several fronts. The relevance of inter- and intra-cellular communication was underlined by the Suda-Nakano groups [32], who wanted to develop a new kind of communication technology – the Molecular Communications – getting inspired by the biological world. The implication of synthetic-to-natural communication and vice versa, in the field of wetware artificial life, was instead emphasized in a perspective paper [7], where a hypothetical Turing test adapted to SCs was firstly discussed. Shortly after, one of the authors of this paper published an experimental report where very simple solute-containing liposomes could produce and release a chemical that was sensed by bacteria [15]. Discussing the future of nanomedicine, LeDuc proposed the term "nano-factories" [20], to indicate programmable SCs that travel into the human body and activate internal circuits for the transformation of metabolites or for the production of a drug. These SCs could recognize ill tissues, communicate with biological cells, and behave consequently.

It seems clear that SCs can indeed be built in order to be able to exchange chemical signals with other SCs or with biological cells. In this way, it would be possible to:

– in general, further advance SC technology
– develop nanomachines for Molecular Communications and bio-chem-ICTs [32]
– create tools for intelligent drug delivery, sensing, diagnosis [19,20]
– investigate what is autonomy and minimal cognition [4]
– put forward a radically new approach to embodied AI [9,10]

Experimental Results

As mentioned, the first example of synthetic communication between cell-like vesicle systems and biological cells was published in 2009 by Ben Davis and collaborators [15]. Lipid vesicles were prepared in order to encapsulate the components of the so-called formose reaction [5,6]; consequently, various types of aldoles were produced and released in the medium. After forming the corresponding adducts with borate (present in the medium) some of these products were able to stimulate a biological response in the bioluminescent bacterium *Vibrio harveyi*. This was possible because *V. harveyi* mistake such artificial products for the true signal molecule AI-2 (autoinducer-2), which has a similar structure and is used by this species for molecular communications.

Based on these pioneering results, we started and promoted by several preliminary publications an experimental program aimed at extending the Ben Davis' approach to SCs, which can be advantageous in terms of design, modularity, programmability [29,38,40,46]. By exploiting cell-free TX-TL systems and enzyme-catalyzed reactions, SCs can indeed produce signal molecules like biological cells. Moreover, they can also receive and decode such types of signal thanks to the incorporation of receptors and/or transcription factors in their inner genetic circuitry. The central idea is that one or more genetic circuits must be inserted inside liposomes in order to produce either the enzyme(s) that synthesize a signal molecule, or the receptor(s) for such molecules.

The resulting SC activity can be regulated, in principle, so that the signal is not sent continuously but only when needed. Analogously, the received signal molecule should trigger a series of steps leading to one (or more) action(s), i.e., operating as actuators do. Interestingly, the association between the signal and the response can be imposed by design, and can differ from what occurs in biological cells. Such a versatility is possible by a careful engineering of molecular parts and devices that are artificially constructed for this specific goal.

Four recent papers have described this sort of SCs approach to molecular communications. In their first work [22], Mansy and collaborators designed SCs acting as 'translators' for *E. coli*. Theophylline only weakly activate bacteria. SCs were constructed in order to sense theophylline and release IPTG, which strongly activates the bacterial response. At this aim, a theophylline riboswitch was employed, generating α-haemolysin (αHL) pores in IPTG-filled vesicles. In the presence of theophylline, SCs synthesized αHL, which allowed the release of IPTG (previously co-entrapped inside liposomes) in the medium. *E. coli* cells uptake IPTG and activate their internal machinery for a response.

The same authors recently reported a two-way chemical communication between SCs and bacteria [21]. In particular, it was shown that SCs, made by encapsulating the proper genetic circuitry inside liposomes, could either be responsive to chemical signals (AHLs), either synthesize and release in the medium these molecules (and thus activating a response in biological cells). Moreover, it was shown that SCs and biological cells could bidirectionally communicate by exchanging chemical signals.

The investigation of Boyden and collaborators [1] demonstrated that, in two separated SCs populations, arabinose (or theophylline in a second example) activate the αHL production, so that pre-encapsulated IPTG (or doxycycline in a second example) could be released, and thus activate a second SC population. This work showed for the first time the establishment of chemical communication between two SCs populations.

The chemical communication between two different types of SCs (vesicles and proteinosomes) was also reported [47]. Stimulated by an AHL, vesicle-based SCs activated the internal TX-TL reactions which produce αHL. The latter creates pore in the lipid membrane, allowing glucose to escape from the SCs into the environment. The released glucose was taken up by enzymes-containing proteinosomes, which sensed this chemical by glucose oxidase/peroxidase coupled reaction path, resulting in the production of a measurable fluorescence.

Our Approach: SCs and *Pseudomonas aeruginosa*

As evident from the above descriptions, *N*-acyl-homoserine lactones (AHLs) are interesting molecules for artificial chemical communications as they are involved in bacterial quorum sensing. We have recently published a report on one-way communication between AHL-producing SCs and *P. aeruginosa* [39].

The system, described in Fig. 2, uses a short chain AHLs, called *N*-butyryl homoserine lactone (C4-HSL) as molecular messenger between the SCs (senders) and the bacterium *P. aeruginosa* (receiver).

Receiver cells need to be "signal-negative", i.e., their capacity of C4-HSL production has been knocked-down by specific genomic deletion. SCs firstly produce the enzyme for C4-HSL production (called RhlI) in well-folded, and thus functioning, form. The enzyme converts two substrates (butyryl-coenzyme A, C4-CoA, and *S*-adenosylmethionine, SAM) into the target molecule C4-HSL. Further details can be found in Fig. 2 caption.

All steps of the SCs mechanism were checked, namely mRNA and RhlI production, C4-HSL identification and quantification, as well as the response of the biological partner (activation of several genes in response to the presence of C4-HSL). The success of molecular communication, both in liquid and gel medium was monitored by bioluminescence or confocal fluorescence microscopy.

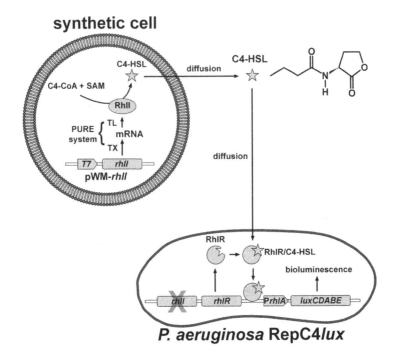

Fig. 2. SCs sending a quorum sensing signal molecule to *P. aeruginosa*. The sending system is based on the production of the signal molecule C4-HSL by the synthase RhlI, encoded by the *rhlI* gene, and two precursors (C4-CoA and SAM). Inside SCs, the RhlI enzyme is produced by the PURE system by transcription (TX) and translation (TL) steps, starting from a DNA template (plasmid pWM-*rhlI*). C4-HSL spreads through lipid membranes, diffuses into the medium and reaches *P. aeruginosa* cells. The receiving cells (named RepC4*lux*) contain a genetic reporter device for C4-HSL-induced bioluminescence (P*rhlA::luxCDABE*) and a mutation inactivating the *rhlI* gene, so that they cannot produce C4-HSL. C4-HSL produced by SCs binds to the receptor RhlR, which in turn triggers *luxCDABE* transcription and bioluminescence emission by RepC4*lux*. Reproduced from [39] with permission from The Royal Society of Chemistry.

4   Implications and Perspectives

The language of information, communication and control theories pervades modern biology. Terms such as code, encoding-decoding, signal, response, receptor, messenger, translation, transcription, and so on, are very common in the vocabulary of any student or professional biologist.

For the first time, however, synthetic biology now provides a conceptual and technical platform for designing and constructing novel information, communication, control systems based on biomolecules and cell-like systems (the SCs). This can be done, importantly, via a bottom-up approach, meaning that the experimentalist knows what are the parts used for building a system, allowing a full control and understanding of its operations.

We would like to shortly comment on two aspects stemming from this nascent research, namely on Molecular Communications and on Minimal Cognition.

Molecular Communications

A totally new branch of wireless communication engineering has been recently launched with the name of "Molecular Communications" [31]. Nakano is one of the most active pioneer in this young field [30,32]. The community of scientists interested in this topic is rapidly growing, and a dedicated series of workshops further witnesses this international trend. Molecular communication technologies aim at exploiting the peculiar features of chemical signals to make systems of nanomachines capable of communicating information, exchanging signals, coordinate their behavior, and so on. This capability of chemical signal manipulation will advance information theory and provide the bases for novel approaches.

However, due to its novelty, almost everything must be built from scratch (experiments and theory), thanks to the collaboration between communication engineers and synthetic biologists. Engineers are interested in theoretical and pragmatic aspects of molecular communications, such as its bandwidth, the noise sources in the emitter and receiver and in the propagation medium, the channel capacity, the probability distribution that best describes a chemical diffusive signal, or the definition of signal-to-noise ratio, and so on [33]. Synthetic biologists are interested in constructing artificial systems capable of *controlled* communication capabilities, so that the resulting systems (i.e., the SCs) are comparable to bio-version of computer and robots. The decodification of information, and the triggered actuation will be central to this enterprise. But, at the same time, this responsiveness will be at the basis of autonomous behavior. This latter consideration brings us to the next fascinating issue.

Minimal Cognition

Wetware artificial models of cognitive processes and systems are rare [16]. Traditionally, cognition has been interpreted either as computation (of heteronomous system) or as self-regulation (of autonomous systems). A detailed discussion on this topic lies outside the scope of this paper, however, we believe that the rise of synthetic biology can radically change the scenario in favor of the self-regulation interpretation, because SCs could be designed in order to perceive changes in their environment and respond accordingly, by modifying their internal dynamical state in a self-regulative manner. In this perspective, the cognitive act is interpreted and realized by the SCs, as a change of their own internal states triggered by a perturbation (the arrival of a signal molecule from the environment). This process is at the basis of biological autonomy [48], and allows the exploitation of SCs as models of basic cognitive systems in the autopoietic sense (i.e., minimal biological cells) [28].

Preliminary discussions on these research directions have been published recently by us [8,10]. In particular, we maintain that SB can contribute to cognitive sciences and artificial intelligence (intended in their minimalistic form) by virtue of its synthetic/constructive paradigm [9]. The goal would be the construction of artificial systems (SCs designed by autopoietic principles) that autonomously *produce* patterns of cognition (for example, the cognitive capabilities of unicellular organisms), by generating, without any symbolic representation, proper self-regulative mechanisms, as living beings do. As observers, we could continue to interpret the absence or the presence of a chemicals in a digitalized 0–1 form, and translate the SCs pattern as it was a computation (e.g., by IF-THEN-ELSE operations). However, SCs would act without the need of a logical central processing unit, but simply by an adaptive response of its reaction network (and being, then, embodied-cognitive).

Acknowledgments. P.S. is grateful to Luisa Damiano (University of Messina, Italy) for inspiring discussions on autonomy, autopoiesis, embodied cognition.

List of Abbreviations

α-HL	α-haemolysin
AI-2	autoinducer-2
AHL	*N*-acyl homoserine lactone
AI	artificial intelligence
C4-CoA	butyryl coenzyme A
C4-HSL	*N*-butyryl homoserine lactone
C.V.	coefficient of variation (standard deviation/mean)
HSL	homoserine lactone
ICTs	information and communication technologies
IPTG	isopropyl β-D-1-thiogalactopyranoside
SAM	*S*-adenosylmethionine
SB	synthetic biology
SC	synthetic cell
TL	translation (from RNA to protein)
TX	transcription (from DNA to RNA)
TX-TL	coupled transcription-translation reactions

Appendix A

With reference to Fig. 1, here we give explanatory comments on the role and relevance of three technologies and on their intersections.

1: Cell-free systems have been traditionally used in biochemical and molecular biology to study biological processes *outside the cell*. Known since decades, the TX-TL systems have been recently re-discovered, also thanks to the introduction of a purified kit whose composition is minimal and perfectly known, the so-called PURE system [44].

2: Liposome technology has been developed mainly for producing liposomes for drug-delivery applications and for biophysical studies. Several methods are available to prepare empty and solute-filled liposomes, depending on the lipid type and vesicle size and morphology. Of particular interest are the so-called 'giant' vesicles (GVs) because they mimic biological cells and can be directly visualized by optical microscopy, for a review on GVs, see [49].

3: Microfluidic devices have been recently introduced in order to manipulate solutions at the micrometer scale. The use of microfluidics for producing water-in-oil droplets is one of the most important application of this technique, but the relevant goal is the production of giant vesicles directly in the microfluidic apparatus (for a review, see [11]). Currently the strategy is based on the formation of water-in-oil-in-water double emulsion droplet, followed by solvent removal.

1–2: These systems are the most used (to date), and consist of liposomes formed by non-microfluidic methods with TX-TL systems (or other biochemical machineries) encapsulated in their aqueous lumen or embedded in their membrane. For a review, see [45]. Note that the populations of spontaneously formed liposomes are generally quite heterogeneous with respect to several parameters (C.V. often >50%).

1–3: Cell-free systems included in microfluidic devices are also a possible combination useful for synthetic biology. For example, work has been carried out to study the dynamics of transcription-translation processes [17].

2–3: To this section belong all work aiming at constructing vesicles, and in particular GVs, by microfluidic devices. It is important to remark that the formation of vesicles in microfluidic devices occurs by repetitive reconstitution of microscopic conditions, and thus all vesicles have very similar structure and a homogeneous population is obtained (C.V. <5–10%). Pioneering work can be found here [27, 36].

1–2–3: This triple overlapping region is probably the 'Holy Grail' of bottom-up SB, namely, the construction by microfluidic devices of solute-filled vesicles, so to have a homogeneous population of bioreactors that can be designed and build according to the general requirements of SB, namely modularity, programmability, reproducibility, etc. An example can be found in [26].

References

1. Adamala, K.P., Martin-Alarcon, D.A., Guthrie-Honea, K.R., Boyden, E.S.: Engineering genetic circuit interactions within and between synthetic minimal cells. Nat. Chem. **9**(5), 431–439 (2017). https://doi.org/10.1038/nchem.2644

2. Altamura, E., Carrara, P., D'Angelo, F., Mavelli, F., Stano, P.: Extrinsic stochastic factors (solute partition) in gene expression inside lipid vesicles and lipid-stabilized water-in-oil droplets. Synth. Biol. (OUP) **3**, ysy011 (2018)

3. Amos, M., Dittrich, P., McCaskill, J., Rasmussen, S.: Biological and chemical information technologies. Procedia Comput. Sci. **7**, 56–60 (2011). https://doi.org/10.1016/j.procs.2011.12.019

4. Bitbol, M., Luisi, P.L.: Autopoiesis with or without cognition: defining life at its edge. J. R. Soc. Interface **1**(1), 99–107 (2004). https://doi.org/10.1098/rsif.2004.0012

5. Breslow, R.: Mechanism of the formose reaction. Tetrahedron Lett. **1**, 22–26 (1959)

6. Butlerow, A.: Bildung einer zuckerartigen Substanz durch Synthese. Annalen **120**, 295–298 (1861)

7. Cronin, L., et al.: The imitation game—a computational chemical approach to recognizing life. Nat. Biotechnol. **24**(10), 1203–1206 (2006). https://doi.org/10.1038/nbt1006-1203

8. Damiano, L., Stano, P.: Synthetic Biology and Artificial Intelligence. Grounding a cross-disciplinary approach to the synthetic exploration of (embodied) cognition. Complex Syst. **27**, 199–228 (2018). https://doi.org/10.25088/ComplexSystems.27.3.199

9. Damiano, L., Kuruma, Y., Stano, P.: What can synthetic biology offer to artificial intelligence (and vice versa)? Biosystems **148**, 1–3 (2016). https://doi.org/10.1016/j.biosystems.2016.09.005

10. Damiano, L., Stano, P.: Understanding embodied cognition by building models of minimal life. In: Pelillo, M., Poli, I., Roli, A., Serra, R., Slanzi, D., Villani, M. (eds.) WIVACE 2017. CCIS, vol. 830, pp. 73–87. Springer, Cham (2018). https://doi.org/10.1007/978-3-319-78658-2_6

11. Elani, Y.: Construction of membrane-bound artificial cells using microfluidics: a new frontier in bottom-up synthetic biology. Biochem. Soc. Trans. **44**(3), 723–730 (2016). https://doi.org/10.1042/BST20160052

12. Filisetti, A., et al.: A stochastic model of autocatalytic reaction networks. Theory Biosci. **131**(2), 85–93 (2012). https://doi.org/10.1007/s12064-011-0136-x

13. Fujiwara, K., Adachi, T., Doi, N.: Artificial cell fermentation as a platform for highly efficient cascade conversion. ACS Synth. Biol. **7**(2), 363–370 (2018). https://doi.org/10.1021/acssynbio.7b00365

14. Furusato, T., Horie, F., Matsubayashi, H.T., Amikura, K., Kuruma, Y., Ueda, T.: De Novo synthesis of basal bacterial cell division proteins FtsZ, FtsA, and ZipA inside giant vesicles. ACS Synth. Biol. **7**(4), 953–961 (2018). https://doi.org/10.1021/acssynbio.7b00350

15. Gardner, P.M., Winzer, K., Davis, B.G.: Sugar synthesis in a protocellular model leads to a cell signalling response in bacteria. Nat. Chem. **1**(5), 377–383 (2009). https://doi.org/10.1038/nchem.296

16. Hanczyc, M.M., Ikegami, T.: Chemical basis for minimal cognition. Artif. Life **16**(3), 233–243 (2010)

17. Karzbrun, E., Tayar, A.M., Noireaux, V., Bar-Ziv, R.H.: Synthetic biology. Programmable on-chip DNA compartments as artificial cells. Science **345**(6198), 829–832 (2014). https://doi.org/10.1126/science.1255550

18. Kita, H., et al.: Replication of genetic information with self-encoded replicase in liposomes. ChemBioChem **9**(15), 2403–2410 (2008). https://doi.org/10.1002/cbic.200800360

19. Krinsky, N., et al.: Synthetic cells synthesize therapeutic proteins inside tumors. Adv. Healthc. Mater. **7**(9), e1701163 (2018). https://doi.org/10.1002/adhm.201701163

20. LeDuc, P.R., et al.: Towards an in vivo biologically inspired nanofactory. Nat. Nanotechnol. **2**(1), 3–7 (2007). https://doi.org/10.1038/nnano.2006.180

21. Lentini, R., et al.: Two-way chemical communication between artificial and natural cells. ACS Central Sci. **3**(2), 117–123 (2017). https://doi.org/10.1021/acscentsci.6b00330

22. Lentini, R., et al.: Integrating artificial with natural cells to translate chemical messages that direct E. coli behaviour. Nat. Commun. **5**, 4012 (2014). https://doi.org/10.1038/ncomms5012

23. Luisi, P.L., Ferri, F., Stano, P.: Approaches to semi-synthetic minimal cells: a review. Naturwissenschaften **93**(1), 1–13 (2006). https://doi.org/10.1007/s00114-005-0056-z

24. Luisi, P.L., Allegretti, M., de Souza, T.P., Steiniger, F., Fahr, A., Stano, P.: Spontaneous protein crowding in liposomes: a new vista for the origin of cellular metabolism. ChemBioChem **11**(14), 1989–1992 (2010). https://doi.org/10.1002/cbic.201000381

25. Majumder, S., Liu, A.P.: Bottom-up synthetic biology: modular design for making artificial platelets. Phys. Biol. **15**(1), 013001 (2017). https://doi.org/10.1088/1478-3975/aa9768

26. Martino, C., et al.: Protein expression, aggregation, and triggered release from polymersomes as artificial cell-like structures. Angew. Chem. Int. Ed. **51**(26), 6416–6420 (2012). https://doi.org/10.1002/anie.201201443

27. Matosevic, S., Paegel, B.M.: Stepwise synthesis of giant unilamellar vesicles on a microfluidic assembly line. J. Am. Chem. Soc. **133**(9), 2798–2800 (2011). https://doi.org/10.1021/ja109137s

28. Maturana, H.R., Varela, F.J.: Autopoiesis and Cognition: The Realization of the Living, 1st edn. D. Reidel Publishing Company, Dordrecht (1980)

29. Mavelli, F., Rampioni, G., Damiano, L., Messina, M., Leoni, L., Stano, P.: Molecular communication technology: general considerations on the use of synthetic cells and some hints from in silico modelling. In: Pizzuti, C., Spezzano, G. (eds.) WIVACE 2014. CCIS, vol. 445, pp. 169–189. Springer, Cham (2014). https://doi.org/10.1007/978-3-319-12745-3_14

30. Nakano, T.: Molecular communication: a 10 year retrospective. IEEE Trans. Mol. Biol. Multi-Scale Commun. **3**(2), 71–78 (2017). https://doi.org/10.1109/TMBMC.2017.2750148

31. Nakano, T., Eckford, A.W., Haraguchi, T.: Molecular Communications. Cambridge University Press, Cambridge (2013)

32. Nakano, T., Moore, M., Enomoto, A., Suda, T.: Molecular communication technology as a biological ICT. In: Sawai, H. (ed.) Biological Functions for Information and Communication Technologies. SCI, vol. 320, pp. 49–86. Springer, Heidelberg (2011). https://doi.org/10.1007/978-3-642-15102-6_2

33. Nanonetworking. http://area51.stackexchange.com/proposals/117498

34. van Nies, P., Westerlaken, I., Blanken, D., Salas, M., Mencía, M., Danelon, C.: Self-replication of DNA by its encoded proteins in liposome-based synthetic cells. Nat. Commun. **9**(1), 1583 (2018). https://doi.org/10.1038/s41467-018-03926-1

35. Oparin, A.I.: The Origin of Life. Traslated by S. Morgulis, 2nd edn. Dover Publications, New York (1953)

36. Ota, S., Yoshizawa, S., Takeuchi, S.: Microfluidic formation of monodisperse, cell-sized, and unilamellar vesicles. Angew. Chem. Int. Ed. **48**(35), 6533–6537 (2009). https://doi.org/10.1002/anie.200902182

37. Pautot, S., Frisken, B.J., Weitz, D.A.: Production of unilamellar vesicles using an inverted emulsion. Langmuir **19**(7), 2870–2879 (2003). https://doi.org/10.1021/la026100v

38. Rampioni, G., Damiano, L., Messina, M., D'Angelo, F., Leoni, L., Stano, P.: Chemical communication between synthetic and natural cells: a possible experimental design. Electron. Proc. Theor. Comput. Sci. **130**, 14–26 (2013). https://doi.org/10.4204/EPTCS.130.4

39. Rampioni, G., et al.: Synthetic cells produce a quorum sensing chemical signal perceived by Pseudomonas aeruginosa. Chem. Commun. **54**, 2090–2093 (2018). https://doi.org/10.1039/C7CC09678J

40. Rampioni, G., et al.: A synthetic biology approach to bio-chem-ICT: first moves towards chemical communication between synthetic and natural cells. Nat. Comput. 1–17 (2014). https://doi.org/10.1007/s11047-014-9425-x

41. Rosenblueth, A., Wiener, N., Bigelow, J.: Behavior, purpose and teleology. Philos. Sci. **10**, 18–24 (1943)

42. Sakakura, T., Nishimura, K., Suzuki, H., Yomo, T.: Statistical analysis of discrete encapsulation of nanomaterials in colloidal capsules. Anal. Methods **4**(6), 1648–1655 (2012). https://doi.org/10.1039/c2ay25105a

43. Sawai, H. (ed.): Biological Functions for Information and Communication Technologies: Theory and Inspiration. SCI. Springer, Heidelberg (2011). https://doi.org/10.1007/978-3-642-15102-6

44. Shimizu, Y., et al.: Cell-free translation reconstituted with purified components. Nat. Biotechnol. **19**(8), 751–755 (2001). https://doi.org/10.1038/90802

45. Stano, P., Carrara, P., Kuruma, Y., de Souza, T.P., Luisi, P.L.: Compartmentalized reactions as a case of soft-matter biotechnology: synthesis of proteins and nucleic acids inside lipid vesicles. J. Mater. Chem. **21**(47), 18887–18902 (2011). https://doi.org/10.1039/C1JM12298C

46. Stano, P., Rampioni, G., Carrara, P., Damiano, L., Leoni, L., Luisi, P.L.: Semisynthetic minimal cells as a tool for biochemical ICT. Biosystems **109**(1), 24–34 (2012). https://doi.org/10.1016/j.biosystems.2012.01.002

47. Tang, T.Y.D., et al.: Gene-mediated chemical communication in synthetic protocell communities. ACS Synth. Biol. **7**, 339–346 (2018). https://doi.org/10.1021/acssynbio.7b00306

48. Varela, F.J.: Principles of Biological Autonomy. The North-Holland Series in General Systems Research. Elsevier North-Holland, Inc., New York (1979)

49. Walde, P., Cosentino, K., Engel, H., Stano, P.: Giant vesicles: preparations and applications. ChemBioChem **11**(7), 848–865 (2010). https://doi.org/10.1002/cbic.201000010

50. Weiss, M., et al.: Sequential bottom-up assembly of mechanically stabilized synthetic cells by microfluidics. Nat. Mater. **17**(1), 89–96 (2018). https://doi.org/10.1038/nmat5005

A Nano Communication System for CTC Detection in Blood Vessels

Luca Felicetti, Mauro Femminella$^{(\boxtimes)}$ (iD), and Gianluca Reali (iD)

Department of Engineering, University of Perugia, CNIT RU,
Via G. Duranti 93, 06125 Perugia, Italy
{luca.felicetti,mauro.femminella,gianluca.reali}@unipg.it

Abstract. In this paper, we show a simulation scenario of a short section of a blood vessel, in which white blood cells, red blood cells, and platelets move as a consequence of collisions and the Hagen–Poiseuille law. In addition to these cells, we have considered also the presence of circulating tumor cells (CTC) and of a receiver node that is able to detect the presence of CTC by using its surface receptors which are affine to the ligands present on the CTC surface.

This study aims at identifying potential optimal positions of CTC sensors within blood vessels in order to maximize the probability of a successful detection.

A simulation campaign has been performed by the BiNS2 simulation framework for several distances of the receiver node from the vessel axis. Obtained results show that CTCs tend to move towards the endothelium.

Keywords: Molecular communications · Simulation ·
Tumor detection · Circulating Tumor Cell · Blood vessels

1 Scientific Background

Nanoscale communications is a new research area that spans on many fields [1,22] and in the recent years lots of studies have focused on the biological and medical cases [5]. This research activity aims to develop a sort of cyber-physical system able to support and improve the natural biological processes against diseases and other degenerative phenomena. In fact, a first step includes the continuous monitoring of the concentrations of specific parameters in the body of the patient and, more in details in the blood stream. This phase could help doctors to detect, monitor and analyze the health conditions, allowing a prompt response increasing the effectiveness of the treatment and reducing, at the same time, undesired side effects.

Supported by the EU project H2020 FET Open CIRCLE (Coordinating European Research on Molecular Communications, project No. 665564) and by the MolML project funded by University of Perugia.

© Springer Nature Switzerland AG 2019
M. Bartoletti et al. (Eds.): CIBB 2017, LNBI 10834, pp. 159–170, 2019.
https://doi.org/10.1007/978-3-030-14160-8_16

A nanoscale communication system composed by tiny devices, called nanomachines, could allow the exchange of health informations from the inner body of the patient to the external, allowing the start of therapies in standard ways and also by means of the release of specific drugs by the nanomachines deployed in the human body. This could give great benefits in the support to the immune system against phatogens but also on the tumor detection and progression inside blood vessels by means the detection of the Circulating Tumor Cells (CTCs).

In healthy conditions a minimal presence of CTCs may be found in the blood stream, and the immune system is typically able to remove them [31]. When a cancer has started to develop, CTCs originate massively from the primary tumor site and, through the bloodstream, may propagate in any part of the body, generating the so-called metastases if suitable conditions are found (Fig. 1). Hence the concentration of such cells in the bloodstream gives useful diagnostic and prognostic information about the location and progression of a tumor. The monitoring of the CTC concentration allows both an early detection of a disease both in its initial phase and any at any relapse of it after an apparently successful treatment.

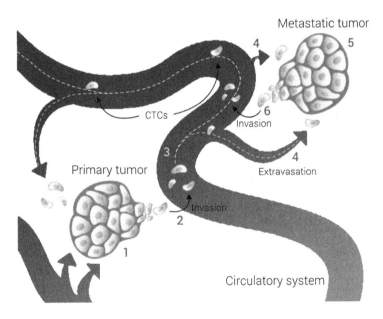

Fig. 1. (1) Circulating Tumor cells detach from the primary tumor site and (2) disseminate into bloodstream (3) so those are virtually capable of reach any part of the body and (4–5) contaminate healthy cells and tissues forming secondary tumor sites (metastasis). (6) The process may restart from the beginning.

Tumors try to evade the immune system by exploiting the regulatory mechanisms that protect healthy cells from immune mediated attacks [6]. Indeed, each CTC exposes several biomarkers on its surface that are useful for the cell detection. One of the most important biomarker is the CD47 that is in general over-expressed by the tumoral cells to fool the immune system and avoid to be destroyed by macrophages [6,28], which are cells having the task of removing dying or dead cells and cellular debris. The CD47 protein, an immunoglobulin (Ig)-like receptor, is normally exposed by many cell types in order to indicate to macrophages that they should not be eliminated. The binding of CD47 expressed on the cell surface with the signal regulatory protein-α (SIRP-α) also known as SHPS1. It acts as a receptor for CD47, thus blocking the phagocytosis of macrophages, as schematically shown in Fig. 2.

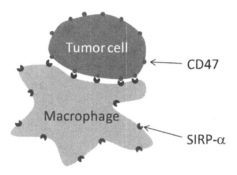

Fig. 2. Tumor cell blocking phagocytosis through CD47 exposure.

For this reason blocking this mechanism through modifying the relationship between CD47 and SIRP-α is an active research area [29]. Current nanomedicine approaches consider RNA interference (RNAi) technology by means of liposomes made with protamine-hyaluronic acid and loaded with anti-CD47 siRNA. This has resulted in an efficient silencing of CD47 and tumor regression [27]. It is worth to mention other important biomarkers, such as the circulating microRNA-101 for the hepatocellular carcinoma [12], the plectin, for the pancreatic cancer [26], the apolipoprotein C-II for the cervical cancer [18], the CD164 protein for the ovarian cancer [20], the plasma osteopontin, for non-small cell lung cancer [17], and the carcinoembryonic antigen (CEA) for different types of lung cancer [14]. All these markers are expressed over the surface of CTCs, which are slightly larger than white blood cells (they seem also to detach from the primary tumors in groups). It may also happen that CTCs spread their DNA throughout the circulatory system by using microvesicles/exosomes as transport vector [8].

Hence, by the simultaneous detection of different types of biomarkers it is possible to identify different cancer types and their stage with a high degree of accuracy, thus obtaining also information about the location of the tumor. It is likely that risk ascertainment methodologies will rely on networks of markers or markers that will undergo remodulation in different progression of the disease.

Table 1. Simulation parameters

General parameters	
Vessel length	6.0 mm
Vessel diameter	30 μm
Mean flow velocity	0.5 mm/s
Viscosity	1.3 mPa·s
Temperature	310 K
Time step	100 μs
Red Blood Cells (RBC)	
Concentration	$5 \cdot 10^6$ U/mm^3
Radius	2.9 μm
White Blood Cells (WBC)	
Concentration	$7.5 \cdot 10^3$ U/mm^3
Radius	3.8 μm
Platelet	
Concentration	$2.5 \cdot 10^5$ U/mm^3
Radius	1.0 μm
CTC and Receiver node	
CTC concentration	10^{-4} U/mm^3
CTC and Receiver radius	5.0 μm
CTC initial distance from axis	[0, 12.5, 25] μm
Initial distance between CTC and Receiver	6 mm

Molecular communications have a central role for realtime detecting CTCs by using implanted devices. Detection can happen by either using contact-based communications [15], or by absorbing microvesicles/exosomes transporting RNA-i strands of the tumor emitted by CTCs [30] in the blood.

The CTCs enumeration could allow the monitoring of the dynamics of the tumor burden and of its spreading to other locations in the body. Moreover, the monitoring phase allows also the detection of any relapse of the disease after a successfull treatment. Hence, the scope of this work is to analyze the distribution of the CTCs along the blood vessels in order to realize an efficient monitoring by means of a smart positioning of the detector nodes.

Nevertheless, the concentration of CTCs in the blood is very low compared to the blood cells concentration and their detection could be quite difficult to achieve [16]. For this reason, in recent years, several research studies have been made in this field and promising results have been achieved by different groups, stimulating further research in the field [13,19,31]. One of the most promising result is the micro-fluidic chip capable to capture CTCs with a high degree of success [4].

2   Simulation Methods

The distribution of the CTCs within blood vessels has been analyzed by means of a simulation campaign performed with the BiNS2 (Biological and Nano-Scale Communication Simulator) simulation framework [9–11].

BiNS2 is an agent-based simulator developed in Java language, able to reproduce, with a high degree of accuracy, the interactions among particles dispersed in a fluid medium inside both unbounded or bounded volumes of different shapes and sizes.

From a communication point of view, some elements can be configured as transmitter or receiver nodes (or both) and other elements as information carriers, implementing some of the typical phases of the communication protocol stack (transmission and reception, signal encoding and decoding, signal modulation and demodulation, information processing, etc.).

BiNS2 allows also the definition of several propagation models, one for each volume, and up to now the most important models defined are: (a) the diffusion-based model, where the Brownian motion affects the propagation of each element dispersed in the medium, and (b) the flow-based model, where each particle is affected by a laminar flow, in addition to a Brownian motion component. The last one is typically used for the simulation of biological environments (i.e. inside blood vessels) or microfluidic devices. BiNS2 accounts explicitly for collisions between different elements, and implements different types of receivers, such as the transparent receiver, the absorbing receiver, and the receiver with absorbing receptors.

The biological scenario used in this study is composed by a short section of a blood vessel in which white blood cells, red blood cells, and platelets move as a consequence of collisions and the Hagen–Poiseuille law [27]. Relevant biological parameters have been set in the MolComML configuration file [2] according to realistic values found in literature [24] and have been reported on Table 1.

As shown by the numerical data, the concentration of the RBCs is some orders of magnitude higher than the concentration of the other cell types [16]. This means that RBCs are predominant in the blood and have an important effect on the rheological properties of the blood. One of the key effect is that these cells tend to aggregate along the central axis of the vessel (characterized by the higher blood speed, due to the Hagen–Poiseuille law) forming a sort of cell-free layer on the section close to the vessel walls. The thickness of this free layer is significant in smaller vessels, whereas it becomes negligible on large vessels [23]. Here, we have assumed an initial cell-free layer comparable to the size of the RBC but the size of the layer may change over the space and time according to the motion law of the particles (mostly due to collisions and Hagen–Poiseuille law).

Assuming to analyze a vessel portion far from the tumor location, the probability of having more than one CTC at time in the same short vessel section is negligible, because their concentration is several order of magnitude lesser than the concentration of the main blood cells [16], as reported on Table 1. Experimental data confirm this assumption as reported in [3,7,21]. Indeed, the authors of [3] show that more than half of the tested patients with metastatic prostate cancer had two or more CTCs per 7.5 mL of blood, and only 18% of them had more than 10 CTCs per milliliter. Since, in our simulation setting the total simulated volume is equal to only 4.24×10^{-6} mL, it is evident that the probability of finding more than one CTCs at time in this blood volume is really rare and unlikely.

We simulate the presence of a CTC at the vessel entrance, and locate both the receiver and CTC at different distances from the vessel axis (i.e. from 0 up to 25 μm). The initial longitudinal distance from the CTC to the receiver node is equal to the total vessel length (6 mm) and, without any loss of generality, the receiver size is set equal to the CTC.

We make the common assumption that the receiver can detect the presence of the CTC by using its surface receptors, which are compliant with the ligands present on the CTC surface (e.g. the CD47 protein) or with any other soluble tumor biomarker that has been released by the tumor cells. Finally, we assume also that the detection phase does not affect the movement of the CTC (hypothesis of transparent receiver).

The noise sources that could affect this communication system are of two types. The first noise component typical of a diffusion-based system (i.e. the *particle sampling noise*) affects the transmitter side [25], but in this scenario there is not any transmitter, because we just count the tumor cells that enter in the observed blood vessel section. Thus, this noise source can be neglected. The second noise contribution is given by the *particle counting noise* [25], which affect the signal propagation. In this case, this contribution is given by the randomness of the Brownian motion contribution on the trajectory of the CTC, as well as by the random displacement due to the collisions with RBCs. Thus, we fully model noise in our simulation scenario.

This study aims at identifying potential optimal positions of a CTC sensor within a blood vessel in order to maximize the CTC detection probability. As said before, detection happens upon collision between the sensor and the CTC, which trigger the chemical reaction of ligands present on the CTC surface with compliant receptors on the receiver one. Thus, the next section shows a comprehensive analysis of the CTC positions as a function of its initial location at the entrance of the blood vessel.

3   Results

Extensive simulations have been performed by means of the BiNS2 simulation framework and the configuration parameters have been setup by the MolComML configuration file. For each case study has been analyzed a different initial position of the CTC cell compared to the distance from the central axis of the vessel. In order to extrapolate the propagation pattern of the CTC, each case study has been simulated at least 10 times.

On Fig. 3 are shown the results for the most representative case studies, for the two extreme positions and for the intermediate position of the CTC, depicted here by a black dot along the thin dotted line on the Top View of each graph.

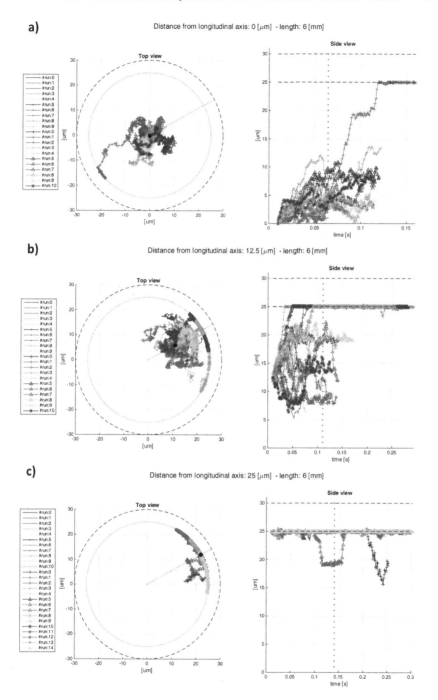

Fig. 3. CTC propagation profile for different distances from the vessel axis: (a) $d = 0\,\mu m$, (b) $d = 12.5\,\mu m$, (c) $d = 25\,\mu m$. Vessel length is $6\,mm$. Different colors are relevant to different simulation runs. (Color figure online)

The circular dotted line on the same view represent the maximum position that could be assumed by the center of the CTC because of the collision with the vessel walls, depicted here by the black dashed line. On the right are shown the side views for each case study and again, the two dashed lines are for the vessel walls and for the maximum allowed position of the CTC. Note that these views show the distance of the CTC from the central axis over the time.

The duration of each run of the same simulation may be different. This is due to the fact that the simulation ends when the CTC crosses the final section of the vessel. The total propagation time depends on the cell velocity, which, in turn, is a function of the actual distance from the central axis, in accordance with the Hagen-Poiseuille law.

Hence, the more the cell is close to the vessel walls and the higher is the crossing time in that section of the vessel.

The results have shown that CTC close to the vessel walls tend to remain close to the endothelium, with a small angular displacement from the initial position, just over $30°$ ($\pi/6$ rad) on both directions, as shown on Fig. 3c.

On Fig. 3b is shown the intermediate case (distance from the axis = $12.5\,\mu$m). The side view show that in most cases the CTC prefer to move towards the vessel walls with respect to the initial position, even if a sort of steady position seems to exists because the CTC is trapped between the blood cells. Anyway there are continuous jumps between different locations due to the collisions with the blood cells and, as shown on the top view, the CTC propagate randomly all around the initial position.

A similar behavior is observed for the last case Fig. 3a. Again, the CTC randomly propagate all around the initial position (top view) but the distance from the central axis (side view) is always increasing (positive slope in the long term for each case). This is due to the fact that the central position is affected by the highest velocity profile and the cell collisions are more energetic and the lightest cells (the CTC in this case) are bounced away by the heavier cells towards the edge of the vessel, where the blood density is lower.

Figure 4a shows the *mean* CTC propagation profile with the relevant 99% confidence intervals for different initial distances from the vessel axis, ranging from: $d = 0\,\mu$m up to $d = 25\,\mu$m. It is evident the trend of CTCs to move away from the center of the vessel, which suggest to position a possible sensor close to the endothelium, which also ensure a possibility to anchor it. Clearly, as shown in Fig. 3, there is also the non negligible possibility that some CTCs will follow a different path, mainly dependent on the sequence of collisions with RBCs. In order to provide a further analysis by considering a sufficiently long blood vessel, these CTC propagation profiles could be sequentially combined, in order to obtain an unique propagation profile given by the succession of the individual profiles (i.e. each profile will starts from the end of the previous one). What emerges from Fig. 4b is that, after a given time (or similarly, after a given distance traveled), *in average* the CTC will tend to drift towards the vessel walls, even if its starting position was near to the central axis of the vessel. Moreover, the peculiarities of the parabolic velocity profile cause a long passing time for

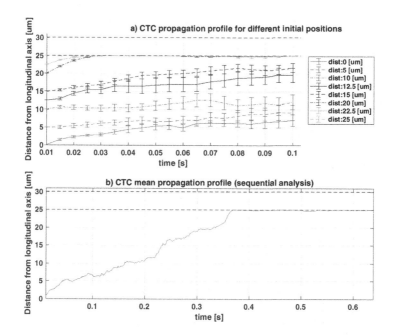

Fig. 4. (a) CTC propagation profile with the relevant 99% confidence intervals for different initial positions from the vessel axis, from: $d = 0\,\mu m$ up to $d = 25\,\mu m$. (b) Sequential analysis of the CTC mean propagation profile.

positions close to the vessel walls, involving long sections of the endothelium to get in touch with the CTC, making reasonable the use of engineered cells on those locations for their detection.

What emerges from these graphs is that in the long term, the CTC seems to move towards the vessel walls regardless to the initial position, with a transverse displacement, around the initial position, which gradually decreases as the cell reaches the edges. Clearly, the area close to the vessel wall is convenient for anchoring engineered cells acting as sensor, but it is also the vastest to monitor. However, as shown in Fig. 3b and c (left sub-figure), once reached the vessel wall, the CTC does not continue its path in a straight longitudinal fashion. Instead, it tends to change its latitude, moving around the vessel wall. This increases the probability that, independently from the initial position of a CTC, a sensor deployed on the vessel wall will have a significant probability to intercept it. The optimal strategy would be engineering portions of the endothelium so as to increase this probability.

These findings are confirmed also by Fig. 5, which shows the mass probabilities versus the distance from the vessel axis for each case study. The position of the CTC are collected when it exits from the simulated section of the vessel. The collected results show that for the first case of Fig. 5a, the positions assumed by the CTC have a sort of Gaussian distribution centered between 0.5 to 0.75 μm (note that the mean value is shifted to the right of the graph, i.e. towards the

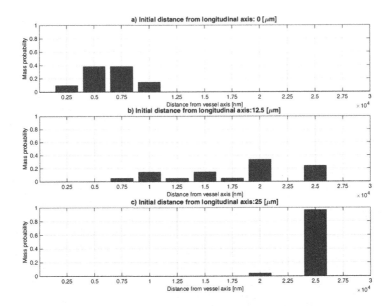

Fig. 5. Distribution of CTC as a function of the distance from the vessel axis

endothelium). Instead, for the intermediate initial position of the CTC (Fig. 5b), we can see that for the half of the cases the position assumed is close to the vessel walls and for the second half the positions are equally scattered towards the center of the vessel. Note also that for this second half, the CTC have spent less time inside the vessel, with respect to the first half, due to higher propagation velocities. Anyway, over than 90% of cases, the CTC have reached a higher position with respect to the initial position, so our opinion is that in the end, for longer vessels, the general trend of the CTCs is to move toward the vessel walls.

Finally, for the last case of Fig. 5c, almost the totality of the CTCs have maintain the initial position (close to the endothelium).

4   Conclusion

In this work we have analyzed the propagation profile of the CTC inside blood vessels for different initial position of the CTCs. The collected results, obtained by a simulation campaign, allow to identify the more convenient location, for the receiver nodes, in order to maximize the probability to intercept CTCs in the blood. A reasonable solution could be to place the CTC sensors on the endothelium, or close to the vessel walls. A lot of feasible solutions could allow the positioning of these detectors, from artificial sensors or stents to modified endothelial cells that could react upon the collision of the CTC.

References

1. Akyildiz, I.F., Jornet, J.M.: The internet of nano-things. Wirel. Commun. **17**(6), 58–63 (2010). https://doi.org/10.1109/MWC.2010.5675779
2. Alarcon, E., et al.: MolComML: the molecular communication markup language. In: Proceedings of the 3rd ACM International Conference on Nanoscale Computing and Communication, NANOCOM 2016, pp. 16:1–16:6. ACM, New York (2016). http://doi.acm.org/10.1145/2967446.2967460
3. Allard, W.J., et al.: Tumor cells circulate in the peripheral blood of all major carcinomas but not in healthy subjects or patients with nonmalignant diseases. Clin. Cancer Res. **10**(20), 6897–6904 (2004)
4. Alunni-Fabbroni, M., Sandri, M.T.: Circulating tumour cells in clinical practice: methods of detection and possible characterization. Methods **50**(4), 289–297 (2010)
5. Atakan, B., Akan, O., Balasubramaniam, S.: Body area nanonetworks with molecular communications in nanomedicine. IEEE Commun. Mag. **50**(1), 28–34 (2012). https://doi.org/10.1109/MCOM.2012.6122529
6. Bordon, Y.: Cracking the combination. Nature Rev. **12** (2013). https://doi.org/10.1038/nri3481
7. Dong, Y., et al.: Microuidics and circulating tumor cells. J. Mol. Diagn. **15**(2), 149–157 (2013). https://doi.org/10.1016/j.jmoldx.2012.09.004. http://www.sciencedirect.com/science/article/pii/S1525157812003078
8. El Andaloussi, S., Mäger, I., Breakefield, X.O., Wood, M.J.A.: Extracellular vesicles: biology and emerging therapeutic opportunities. Nature Rev. Drug Disc. **12**, 347 EP (2013). https://doi.org/10.1038/nrd3978, perspective
9. Felicetti, L., Femminella, M., Reali, G.: A simulation tool for nanoscale biological networks. Nano Commun. Netw. **3**(1), 2–18 (2012)
10. Felicetti, L., Femminella, M., Reali, G.: Simulation of molecular signaling in blood vessels: software design and application to atherogenesis. Nano Commun. Netw. **4**(3), 98–119 (2013)
11. Felicetti, L., Femminella, M., Reali, G., Gresele, P., Malvestiti, M.: Simulating an in vitro experiment on nanoscale communications by using BiNS2. Nano Commun. Netw. **4**(4), 172–180 (2013)
12. Fu, Y., et al.: Circulating microRNA-101 as a potential biomarker for hepatitis B virusrelated hepatocellular carcinoma. Oncol. Lett. **6**(6), 1811–1815 (2013). https://doi.org/10.1038/nri3481
13. Gidaspow, D., Huang, J.: Kinetic theory based model for blood flow and its viscosity. Ann. Biomed. Eng. **37**(8), 1534–1545 (2009)
14. Grunnet, M., Sørensen, J.: Carcinoembryonic antigen (CEA) as tumor marker in lung cancer. Lung Cancer **76**, 138–143 (2011)
15. Guney, A., Atakan, B., Akan, O.: Mobile ad hoc nanonetworks with collision-based molecular communication. IEEE Trans. Mob. Comput. **11**(3), 353–366 (2012). https://doi.org/10.1109/TMC.2011.53 ·
16. Haber, D.A., Velculescu, V.E.: Blood-based analyses of cancer: circulating tumor cells and circulating tumor DNA. Cancer Discov. **4**(6), 650–661 (2014)
17. Han, S.S., et al.: Plasma osteopontin is a useful diagnostic biomarker for advanced non-small cell lung cancer. Tuberc. Respir. Dis. **75**(3), 104–110 (2013)
18. Harima, Y., et al.: Apolipoprotein C-II is a potential serum biomarker as a prognostic factor of locally advanced cervical cancer after chemoradiation therapy. Int. J. Radiat. Oncol. Biol. Phys. **87**, 1155–1161 (2013)

19. Hsieh, H.B., et al.: High speed detection of circulating tumor cells. Biosens. Bioelectron. **21**(10), 1893–1899 (2006)

20. Huang, A.F., Chen, E., Huang, S.M., Kao, C.L., Lai, H.C., Chan, J.Y.-H.: CD164 regulates the tumorigenesis of ovarian surface epithelial cells through the SDF-1/CXCR4 axis. Mol. Cancer **12**, 115 (2013)

21. Kowalik, A., Kowalewska, M., Gozdz, S.: Current approaches for avoiding the limitations of circulating tumor cells detection methods-implications for diagnosis and treatment of patients with solid tumors. Transl. Res. **185**, 58–84 (2017)

22. Nakano, T., Moore, M., Wei, F., Vasilakos, A., Shuai, J.: Molecular communication and networking: opportunities and challenges. IEEE Trans. Nanobiosci. **11**(2), 135–148 (2012). https://doi.org/10.1109/TNB.2012.2191570

23. Pan, W., Caswell, B., Karniadakis, G.E.: A low-dimensional model for the red blood cell. Soft Matter **6**(18), 4366–4376 (2010)

24. Phillips, K.G., et al.: Optical quantification of cellular mass, volume, and density of circulating tumor cells identified in an ovarian cancer patient. Front. Oncol. **2**, 72 (2012)

25. Pierobon, M., Akyildiz, I.F.: Diffusion-based noise analysis for molecular communication in nanonetworks. IEEE Trans. Signal Process. **59**(6), 2532–2547 (2011). https://doi.org/10.1109/TSP.2011.2114656

26. Shin, S., et al.: Unexpected gain of function for the scaffolding protein plectin due to mislocalization in pancreatic cancer. Proc. Natl. Acad. Sci. **110**, 19414–19419 (2013)

27. Tan, J., Thomas, A., Liu, Y.: Influence of red blood cells on nanoparticle targeted delivery in microcirculation. Soft Matter **8**, 1934–1946 (2012)

28. Unanue, E.R.: Perspectives on anti-cd47 antibody treatment for experimental cancer. Proc. Nat. Acad. Sci. **110**(27), 10886–10887 (2013). https://doi.org/10.1073/pnas.1308463110. http://www.pnas.org/content/110/27/10886

29. Weiskopf, K., et al.: Engineered SIRP variants as immunotherapeutic adjuvants to anticancer antibodies. Science **341**, 88–91 (2013)

30. Xie, Z., Wroblewska, L., Prochazka, L., Weiss, R., Benenson, Y.: Multiinput RNAi-based logic circuit for identification of specific cancer cells. Science **333**(6047), 1307–1311 (2011). https://doi.org/10.1126/science.1205527. http://science.sciencemag.org/content/333/6047/1307

31. Yu, M., Stott, S., Toner, M., Maheswaran, S., Haber, D.A.: Circulating tumor cells: approaches to isolation and characterization. J. Cell Biol. **192**(3), 373–382 (2011)

Experimental Evidences Suggest High Between-Vesicle Diversity of Artificial Vesicle Populations: Results, Models and Implications

Pasquale Stano[1](✉) , Roberto Marangoni[2,3] , and Fabio Mavelli[4]

[1] Department of Biological and Environmental Sciences
and Technologies (DiSTeBA), University of Salento, Ecotekne, 73100 Lecce, Italy
pasquale.stano@unisalento.it
[2] Department of Biology, University of Pisa, Via Derna 1, 56126 Pisa, Italy
[3] CNR – Institute of Biophysics, Via G. Moruzzi 1, 56124 Pisa, Italy
[4] Department of Chemistry, University of Bari, Via Orabona 4, 70125 Bari, Italy

Abstract. In the past years, artificial cellular models for origins-of-life research and synthetic biology have been extensively studied. At this aim, solute-filled lipid vesicles (liposomes) are widely used. Several evidences have been collected about the capture of water-soluble chemicals, the mechanism of vesicle self-reproduction, and the course of (bio)chemical reactions in the vesicle lumen. Among the several fascinating questions which emerged from these studies, here we focus on a peculiar feature, namely, the fact that a spontaneous heterogeneity of vesicle structure often emerges. In other words, vesicle populations created in the laboratory by classical batch methods include very 'diverse' vesicles with respect to size, morphology, and – importantly – solute content. The consequences of this between-vesicle diversity are shortly discussed.

Keywords: Autopoiesis · Lipid vesicles · Synthetic biology · Synthetic cells · Primitive cells · Origins of life

1 Lipid Vesicles in Origin-of-Life Research and in Synthetic Biology

Lipid vesicles (liposomes) are versatile microscopic structures originated from the self-assembly of lipids. They are hollow aqueous micro-compartments, often spherical, whose boundary is a semi-permeable membrane composed by two juxtaposed lipid layers (the so-called 'bilayer'), where lipids are arranged tail-to-tail. Lipid bilayers form spontaneously as soon as lipids are dissolved in an aqueous solution, and normally they bend and self-seal to form a closed spherical compartment (typical diameters from ca. 30 nm to ca. 100 μm). Water-soluble compounds, if present in the aqueous phase, become passively encapsulated in the

M. Bartoletti et al. (Eds.): CIBB 2017, LNBI 10834, pp. 171–185, 2019.
https://doi.org/10.1007/978-3-030-14160-8_17

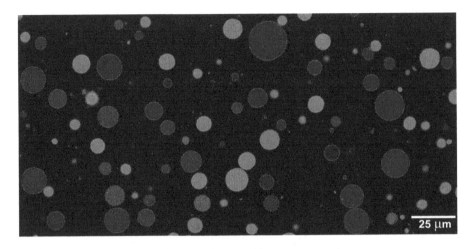

Fig. 1. Giant lipid vesicles are often used for constructing cell-like systems. The picture shows calcein-filled vesicles whose membranes have been stained by Trypan Blue. Reproduced from [38] according to the CC-BY license. (Color figure online)

vesicle lumen, whereas hydrophobic and amphiphilic chemicals will self-localize in the membrane (Fig. 1).

The vesicle formation is spontaneous. This is due to entropic factors, and it represents a very remarkable exergonic mechanism for achieving highly ordered structures (the lipid bilayer, and thus the vesicle) at the expenses of the water molecules of the solvent, whose entropy increases. Moreover, the bilayer closure to form a vesicle further dissymmetrizes the system by generating a *distinction* between the intra-vesicle chemical composition and the external one.

Due to these features, lipid vesicles are used as simplified models of biological cells in two apparently diverse research areas, namely origins of life and synthetic biology. In origins of life investigations, lipid vesicles are taken as primitive cell models. The 'constructive' approach [6,17] represents the unique way to study primitive cells – so to explore their structure, behavior, stability, reactivity, capacity of encapsulating other substances, and so on. On the other hand, the same approach can be adopted in synthetic biology to build simplified cellular models aiming at the engineering and construction of cell-like structures in controlled and programmed way. These 'minimal synthetic cells' can help understanding of how molecular biosystems work or can accomplish specific biotechnological tasks. As mentioned, these two research directions, which might appear almost at the antipodes, share as their common ground the theory and the practice of *constructing* cellular models in the laboratory [37].

2   Inquiring into the Cell Cycle of Primitive Cells

Experimental research on vesicles as primitive cell models was started in the 1990s essentially by the group of Pier Luigi Luisi at the ETH Zürich. Prior models were indeed based on coacervates [27] or microspheres [10] (see [13] for an historical review), but the discovery of liposomes by Alec Bangham [2] and especially the several studies on fatty acid vesicles [11,15] strongly promoted the shift toward vesicles as biomimetic compartments.

Two elements of innovation were introduced by the Swiss group. Firstly, the target was the assembly of enzyme-filled vesicles (analogous to the Oparin enzyme-filled coacervates). Second, the vesicles were not static, but thanks to the incorporation of new membrane building blocks, they could grow and divide in order to simulate a protocellular proliferation. In other words, from the very beginning it was clear that primitive cell models have to incorporate two of the main features of living cells: an internal reaction network and the grow of the 'shell', and ideally perform a sort of primitive cell cycle, as shown in Fig. 2.

The two chemical systems (the 'core' and the 'shell') should act harmonically so to produce – inside the compartment – all constituents of the primitive cells. Only in this way the daughter cells will be able to re-start the life cycle.

To date, a vesicle system capable of performing a primitive cell cycle as that one depicted in Fig. 2 has *not* been realized yet. It is worth mentioning that such a system would be *autopoietic* [23], as it would be able to autonomously produce all its internal components from internal reactions. However several experimental studies have inquired into the several physico-chemical processes underlying that complex pattern.

Commenting on such results lies outside the scope of this paper. The interested readers can find a detailed account in a recently published review [35]. For the present discussion it is enough to mention that great efforts have been made (and are still made) to study internalized reactions either by employing allegedly primitive compounds (ribozymes, short peptides), either 'modern' molecules (protein, ribosomes, DNA). Ideally, an internal reaction network, irrespective from its material constitution, in order to be autopoietic should be able to synthesize all components of the network, and the boundary compounds as well. For the latter goal, lipid synthesis inside vesicles can be achieved [8,19,30] but not to the point of observing growth and division. Vice versa, if excess boundary-forming compounds (or their precursors) are added from outside, vesicle self-reproduction has been demonstrated [41,44]. Moreover, it is of great interest the physical understanding of the 0th-step of the cell cycle in Fig. 2, namely, the 'booting' self-assembly process that leads to a solute-containing vesicle from separated components.

3   The Need of a Population Perspective: Vesicle 'Diversity'

Overlaid to the dynamics illustrated in Fig. 2, which refers to the molecular and supra-molecular processes of chemical self-assembly, entrapment, transformation, growth-and-division, there is a central physical aspect that was neglected until a few years ago. This is the inescapable *between-vesicle variability* that emerges from the microscopic nature of these assemblies. The interplay between

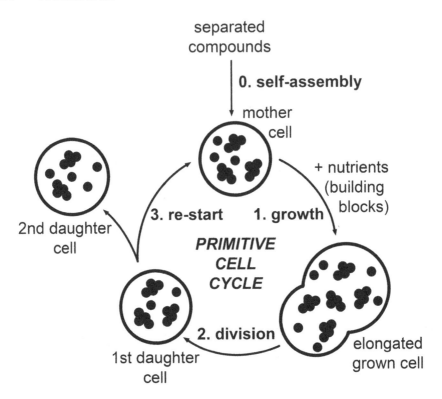

Fig. 2. The hypothetical cycle of a primitive cell. 0th step: formation of primitive cell by lipid self-assembly and solute capture; 1st step: primitive cell growth at the expenses of external compounds, which should enter into the primitive cell (e.g., by diffusion, by the help of a primitive translocator), and being processed by the internal metabolic network so to produce primitive cell own components, including the membrane, so to allow a growth; 2nd step: division of the grown mother cell into two or more daughter cell, a process that implies solute partition; 3rd step: each one of the daughter cell re-start the cycle, giving rise to primitive cell proliferation. The transformations here depicted independent from the chemical nature of primitive cell components, but obey to an autopoietic mechanism whereby the internalized reaction network produces all its components and the boundary molecules as well. Note that an autopoietic mechanism is a prerequisite for a cell cycle, but does not imply it (an autopoietic cell can stay in a stationary homeostatic state without self-reproduction, but still displaying – as internal activity – a continuous production and degradation of its components, membrane ones included). For a discussion on autopoiesis, and reported experiments on artificial autopoietic self-reproduction and autopoietic homeostasis, see, respectively [21,41,43].

their small volume and the stochastic events at the molecular scale originate this additional level of description, that we refer as vesicle 'diversity' [39].

Although it was self-evident that a population of vesicles – intended as cell models – is clearly composed by different vesicles, because of different size, different content, different membrane composition, different number of lamellae, different metabolic efficiency, and so on, an adequate level of understanding of this diversity was not available.

Referring again to Fig. 2, vesicle diversity is generated and associated at each step of the cycle and its booting. In particular, at the 0th step, vesicle self-assembly and solute encapsulation generate probably the largest differences, but also the cycle steps such as the growth (due to diverse reaction rates of the network reactions), and the division (when compartmentalized solutes of the grown mother vesicle are distributed among daughter cells).

Each one of the above-mentioned events might generate large between-vesicle structural diversity, which might translate into functional diversity, and thus generating competition between vesicles (between protocells) - a typical 'biological' behavior. Moreover, the existence and the observation of diversity and of 'extremes' in vesicle populations can be highly informative about the mechanisms of life origins. This diversity is actually a faithful representation of primordial protocellular systems, where populations of lipid compartments spontaneously emerged from the multi-molecular *milieu* created by a contingent combination of local physico-chemical factors.

On the other hand, it should be noted that when vesicles are employed as synthetic cells (in the biotechnological or synthetic biology arenas), between-vesicle diversity should be kept at the lowest possible level, so to have homogeneous populations where all individuals behave the same manner.

4   The Numbers of Vesicle Diversity

Among the various sources of between-vesicle diversity, the number and the type of solutes encapsulated in their lumen provides the most striking effect of diverse functionality in these cell-like systems.

All steps of the primitive cell cycle (Fig. 2) influence (and are influenced by) the composition of the internal vesicle solution as specified in Table 1.

Table 1. Sources and effects of between-vesicle diversity in a primitive cell cycle

Process	Source of diversity	Diversity production
Self-assembly	size, lamellarity, solute content	composition, concentration, S/V ratio
Growth	binding, kinetics, intrinsic stochastic effects	reaction feasibility, performance
Division	size, solute partition among the daughter cells	reactions of next generation

In the past few years we have explored the diversity of vesicle population with respect to the capture of solutes (0th step: bilayer self-assembly and vesicle formation) and partition of solutes among daughter vesicles (3rd step: vesicle division).

The two cases are somehow similar because the solute capture and solute partition can be modeled by the same maths. In particular, we defined a simple model based on a very reasonable 'null hypothesis' H_0, which states that when lipids self-assemble into a vesicle, it is expected that the average number λ of solutes captured in the vesicle lumen is proportional to the vesicle volume. This is equivalent to say that each vesicle 'samples' the aqueous solution where it forms, and capture the solutes dissolved in it, with a probability $p = v/V_{tot}$, where v is the vesicle volume and V_{tot} is the total aqueous volume. Because there are N_{tot} molecules in V_{tot} (thus $C_{bulk} = N_{tot}/V_{tot}$, then $\lambda = p \times N_{tot}$ and $c_{vesicle} = C_{bulk}$.

It follows that λ can be easily calculated as follows:

$$\frac{\lambda/N_A}{v} = c_{vesicle} = C_{bulk} \tag{1}$$

$$\lambda = N_A \cdot C_{bulk} \cdot v \tag{2}$$

where N_A is the Avogadro's number. For example, if $C_{bulk} = 5\,\mu M$, λ is 12.6 and $\sim 2 \times 10^5$, respectively, for vesicles with diameter of 200 nm and of 5 μm.

Owing to stochastic fluctuations of the local (microscopic) C_{bulk} near the nascent vesicles, it is expected that the *actual number* n of entrapped solute molecules differs from λ (and $c_{vesicle}$, the actual intra-vesicle solute concentration, which is proportional to n, will differ from C_{bulk}). As order of magnitude, it is expected that $\Delta\lambda/\lambda$ goes as $\lambda^{-1/2}$ (as obtained by the fluctuations theory). Given a certain C_{bulk}, high variance should be observed for small vesicles, and small variance for large vesicles. Referring to the above-mentioned cases, $\Delta\lambda/\lambda$ should be $\sim 30\%$ and 0.2%, respectively.

As mentioned, according to H_0, the entrapment process is equivalent to random sampling the solution by the nascent vesicle. Due to stochastic fluctuation of the solute concentration in the solution volume, a vesicle will capture n solutes when instead λ are expected on average. It follows that the number of entrapped solute in the vesicle population is distributed according to the Poisson distribution (Eq. 3):

$$p(n) = \frac{\lambda^n}{n!}e^{-\lambda} \quad (n = 0, 1, ...) \tag{3}$$

where λ is the mean value, and $\sqrt{\lambda}$ is the standard deviation. When λ is high, i.e., >10–15 (which in turn means large C_{bulk} and/or large V), the Poisson distribution becomes similar to the Normal (Gaussian) distribution.

However, the same mathematics can be applied to the question of solute partition during the division of a grown mother vesicle (2nd step in Fig. 2). If the distribution of solutes in the mother vesicle is spatially homogeneous, the expected number of solutes that will be found in the daughter cells will be proportional to their volume. Again, the λ value can be estimated by Eq. 2 and the distribution by Eq. 3.

The discussion becomes more complicated when solutes of different types (different chemical species) are considered. The simplest hypothesis is that the entrapment of a species does not depend on the presence of the others, so that the co-entrapment probability is simply given by the product of each entrapment probability (of each individual solute). In mathematical terms (for just two solutes A and B):

$$p(n_A \text{ and } n_B) = p(n_A) \cdot p(n_B) \tag{4}$$

In general, for k different species, it is the product of k Poissonian terms with averages $\lambda_k = N_A \cdot C_{k,\text{bulk}} \cdot v$ [16,18,31].

5   Experimental Evidences

There are not many studies specifically dealing with the question of between-vesicle diversity associated with the vesicle formation, and rare are those dealing with the steps of primitive cell cycle.

In a recent review [1], we have commented quite in detail the available experimental results. Only few authors have studied the solute occupancy distribution inside conventional and giant vesicles in the simple case of one-solute system. At this aim, non-averaging techniques are required, such as microscopy or flow cytometry. On the contrary, numerous reports are based on multiple-solute systems, but the studies do not address the measurement of the solutes concentration, rather focus on the outcome of complex reactions inside vesicles, such as the transcription-translation (TX-TL). TX-TL reactions lead to the production of a protein – a keystone step for this synthetic biology branch.

Anyway, most investigations, but not all, indicate that the between-vesicle diversity is often high, beyond the statistical expectations.

Let us summarize here a series of results, obtained by us, dealing with 0th and 2nd step of the primitive cell cycle (Fig. 2).

5.1   Spontaneous Formation of Solute-Filled Vesicles

Among the four traditional methods for vesicle preparation, the thin lipid film hydration ('natural swelling') is certainly the aptest for simulating the spontaneous formation of primitive cell-like structures. It consists in the deposition of a thin lipid film over a surface (e.g., in a glass test tube) followed by a lipid swelling in the presence of a solution of interest.[1]

It is probably evident to most liposomists that vesicles formed by the thin lipid film hydration method are diverse in terms of morphology, size, lamellarity and – especially – solute content. No detailed discussions have been reported

---

[1] Ideally, one would like to have a single lipid monolayer deposited over a large surface so that all parts of the film would experience the same conditions. This corresponds, in most cases, to work well below the μM lipid concentration range, with consequent vesicle losses and other impractical complications. Thus, in the most common experimental conditions the film is rarely so perfect and different regions of the film will experience different micro-environments. Actually, realistic laboratory conditions might affect the measured heterogeneity of vesicle formation paths.

on the reasons for the latter heterogeneity (the width of the solute occupancy distribution) and on the expected-versus-observed variance.

We studied the encapsulation of ferritin, ribosomes and ribo-peptidic complexes (reviewed in [32]) in conventional phospholipid vesicles prepared by the film method. These are convenient macromolecular solutes – especially ferritin – because they can be visualized by cryo-transmission electronmicroscopy, and directly counted. Thus, the intravesicle solute occupancy distribution can be determined by analyzing a large number of images. Our expectation was to find the classical bell-shaped distribution (Poisson or Gaussian), but we surprisingly found a power-law distribution (Fig. 3) and interesting 'super-filled' vesicles whose formation cannot be easily explained by stochastic fluctuations.

According to these observation, that have been confirmed qualitatively (but not quantitatively) when fluorescent proteins were encapsulated inside small (1–2 μm) giant vesicles [5,39], the actual solute occupancy distribution might vary significantly from expectations, and in particular the formation of solute-filled vesicles, although rare, can shed light on the self-organization drive to the onset of early cells.

When two or more solutes are considered, Eq. 4 extended to the $k = 80$ different solutes of TX-TL reaction has shown limitations, because we recorded again solute rich vesicles even if statistically their occurrence should have been extremely rare. The solute-rich vesicles were those capable of synthesizing a protein. This was proved by running experiments in those conditions that give small λ_k values in two different ways, namely, using small vesicles (small v) [31], or using small $C_{k,\text{bulk}}$ [36]. When conventional vesicles, with diameter of ca. 200 nm have been used, it results that although most vesicles cannot synthesize proteins, some still can, despite the very unfavorable co-entrapment probability (of the order of 10^{-26}). It has been shown that a power-law model can successfully explain the observation of TX-TL reactions in small GVs [25].

An attempt to explain the physics of solute entrapment and the reported power law [22] has been proposed, based on the Cox's theory of renewal point processes [28], but a definitive clarification is still missing.

5.2   Division of a Grown Mother Vesicle into Two or More Daughter Vesicles

In order to model the spontaneous division of primitive cells we devised a mechanical approach whereby solute-filled phospholipid vesicles were fragmented in many small daughter by a mechanical non-spontaneous process called extrusion [9]. Although this strategy clearly represents a crude approximation of the spontaneous vesicle self-reproduction [41,44], it includes the core mechanism of interest for our discussion, i.e., the stochastic partition of encapsulated solutes in the daughter vesicles. In the past, extrusion was applied to primitive cell models, providing information about *average* solute retention [14,34].

We investigated how a population of giant vesicles (average diameter 5–8 μm) behave when vesicles are extruded to give small (0.8 μm) daughter vesicles. The study was carried out by firstly encapsulating fluorescent solutes of different size

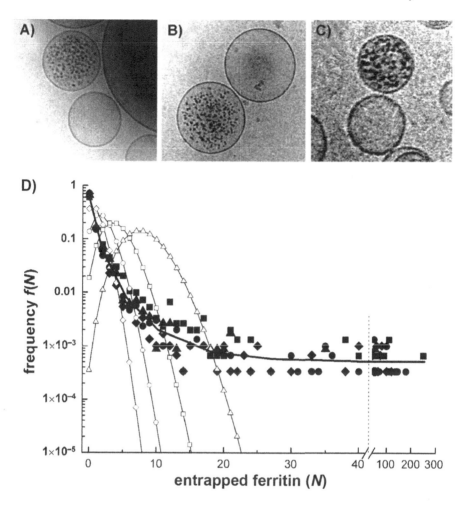

Fig. 3. Encapsulation of ferritin inside conventional lipid vesicles. (A-B-C) Typical super-filled vesicles which contain a higher-than-expected number of ferritin molecules. Such special vesicles are present in minor amount (ca. 0.1%). In particular, the intravesicle ferritin concentration is 8.5× (A), 3.8× (B), 11.8× (C) the expected bulk concentration (C_{bulk}, *cf.* Eq. 2). In the bottom panel (D), the expected (empty symbols) and the measured (filled symbols) solute occupancy distribution is reported. Note the logarithmic scale on the y-axis. The experimental data, when plotted on a double logarithmic plot, give a straight line (a power-law distribution). Reproduced from [22] with the permission of Wiley.

inside giant vesicles, then by extruding giant vesicles and measuring the internal solute concentration of the daughter vesicles (by confocal microscopy). According to H_0 the solute concentration in the mother vesicles and in the daughter ones should be the same, although the spread around the mean value can differ due to stochastic fluctuations.

Figure 4, taken from [9], summarizes the experimental results. In particular, the plots compare the experimental solute occupancy distribution with the theoretical one, obtained by stochastic partition of the solutes initially present in giant vesicles. In this manner not only the average but also the variance of the expected and measured distributions is compared.

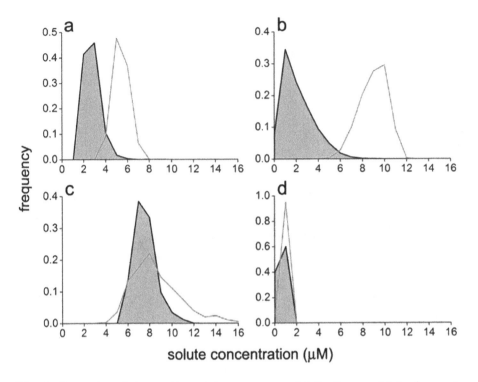

Fig. 4. Comparison between the solute occupancy distribution as obtained experimentally (grey area) and by numerical modeling (red lines). The distributions refer to extruded filled with (a) pyranine; (b) calcein; (c) fluorescein-marked bovine serum albumin; (d) fluorescein-marked dextran. Reproduced from [9] with permission from The Royal Society of Chemistry. (Color figure online)

Results show that the solute partition pattern depends on the solute size. Small solutes such as pyranine (0.52 kDa) and calcein (0.62 kDa) are largely lost during the extrusion of giant vesicles (−48% and −76%, respectively).[2] In contrary, large solutes such as bovine serum albumin (69 kDa) and dextran (150 kDa) – both marked by fluorescein) – experience a lower lost (−11% and −30%, respectively).

---

[2] Note that this loss does not refer to the volume loss which follows from the vesicle size reduction, but it is an authentic concentration reduction due to the reduction of the average number per unit of volume.

Data have been explained on the basis of the different diffusion coefficient (D) of small and large solutes. The high D of small solutes implies a greater movement in the time interval of $\sim 1\,ms$ when the daughter vesicles seal their membrane (after being pinched off from the mother vesicle). In most cases then, the solutes will diffuse away from the volume captured by the daughter vesicle. Vice versa, large solutes will mainly stay and be captured due to their low D.

6 Relevance, Future Challenges, and the Role of Bioinformatics

In this work we have emphasized the role of vesicle diversity as derived from spontaneous processes such as vesicle formation, growth, division. We have shown that when this aspect is specifically investigated, interesting observations are collected – but most of them are currently without a clear explanation.

What are the very generative mechanisms and what is the role of vesicle diversity when primitive cell models or synthetic cells are built in the laboratory?

While undergoing and future research tools need to be sharpened in order to solve the first question, it is already evident that vesicle diversity is a conceptually important factor to consider. Focusing on this topic represents already a cultural shift that can lead to interesting developments. For example, the emergence of primitive cells, often discussed from an individualist viewpoint should be instead discussed from a *population* perspective, and the factors that amplify or propagate any between-individual variations should be taken into account.

The new vision based on protocell ensembles makes the story more interesting. The duality between competition and cooperation, two central ideas in biology, makes sense only when structural and functional diversities are explicitly considered. In the prebiotic context, such a diversity is strongly linked to structure and precedes the onset of genotype variations. The structural difference between protocells is more radical as it originates from the mechanisms of vesicle formation (by self-assembly or by division of a parent vesicle). Therefore, protocell diversity has realistically impacted on early pre-biological evolution.

Protocell assemblies models have been already put forward [4,12,33] but more research is needed. Vesicles could be included in a gel matrix [29] so to create large tissue-like models.

Synthetic biology applications, on the other hand, require highly homogeneous samples. Although most of published work generally rely on batch vesicle preparation methods (and thus display high between-vesicle variability), modern trends suggest that the microfluidic technologies [7,40,42] will revolution the experimental approaches to synthetic cells. We expect indeed that synthetic cell construction by microfluidics will overcome soon the traditional batch methods [35].

The evidences here summarized strongly call for integrate theoretical, experimental, and stochastic modeling approaches [3,9,20,24,25].

Bioinformatics can be an important tool for understanding the behavior of vesicle populations, and realize a qualitative jump in the direction of a systemic and multi-scale view. Bioinformatics can be introduced at three different levels.

1. In the theory of minimal living systems, the conceptualization of the so-called "organizational closure", which characterize autopoietic systems, has been central. Now that synthetic approaches have become realistic, organizational closure can be designed by bioinformatics. What is the minimal complexity of a reaction network capable of self-sustainment? How to understand whether the production of certain component(s) can be omitted from the autopoietic system and instead be 'outsourced' to the environment? What are the network topology features that define autonomy?

2. As modeling support for the experimental approaches, numerical simulations can be very useful in two important aspects that determine the course of micro-compartmentalized reactions, namely extrinsic and intrinsic stochastic effects. The phenomena discussed in Sect. 5 are based on extrinsic stochastic effects. Reactions inside microcompartments, however, are subjected to relevant intrinsic stochastic effects, due to the small number of molecules confined in the vesicle lumen or in the membrane. This fact implies that the reactions occurring in primitive cell models and synthetic cells are best simulated by stochastic kinetics (i.e., master equations, Monte-Carlo methods, etc.), rather than with ordinary differential equations.

3. In the emerging field of bio-chemical-based Information Technologies (bio-chem-ITs) and Molecular Communications [26], bioinformatics becomes functional to develop synthetic cells capable of manipulating chemical signals in terms of the communication theory, from relatively simple systems such as signal exchanges between synthetic cells, to coordinated or synchronized behavior of synthetic cell assemblies, up to more complex pattern as can be a rudimentary differentiation – which are experimental goals probably not so distant in the future.

7   Concluding Remarks

In this paper we have emphasized and discussed the important but often neglected topic of between-vesicle diversity, referred to the current experimental approaches to primitive cell models and synthetic cells. Experiments show that experimentally obtained vesicle populations display quite high variances (except when new microfluidics methods are employed). Accordingly, the definition of an 'average behavior' might become almost meaningless. But instead of being a bad news, these evidences offer an opportunity for better understanding the generative mechanism of protocells, the paths that lead to prebiotic evolution, and the early cell communities.

Finally, we remark that as laboratory-made artificial cells are very simple systems when compared to biological living cells, the application of rigorous bioinformatic models is not only helpful, but also very meaningful.

Acknowledgments. Collaboration among the authors has been fostered by the European COST Action CM1304 *Emergence and Evolution of Complex Chemical Systems.*

References

1. Altamura, E., Carrara, P., D'Angelo, F., Mavelli, F., Stano, P.: Extrinsic stochastic factors (solute partition) in gene expression inside lipid vesicles and lipid-stabilized water-in-oil droplets. Synth. Biol. (OUP) p. (2018, in press)
2. Bangham, A.D., Horne, R.W.: Negative staining of phospholipids and their structural modification by surface-active agents as observed in the electron microscope. J. Mol. Biol. **8**, 660–668 (1964)
3. Calviello, L., Stano, P., Mavelli, F., Luisi, P.L., Marangoni, R.: Quasi-cellular systems: stochastic simulation analysis at nanoscale range. BMC Bioinf. **14**, S7 (2013). https://doi.org/10.1186/1471-2105-14-S7-S7
4. Carrara, P., Stano, P., Luisi, P.L.: Giant vesicles "colonies": a model for primitive cell communities. Chembiochem Eur. J. Chem. Biol. **13**(10), 1497–1502 (2012). https://doi.org/10.1002/cbic.201200133
5. D'Aguanno, E., Altamura, E., Mavelli, F., Fahr, A., Stano, P., Luisi, P.L.: Physical routes to primitive cells: an experimental model based on the spontaneous entrapment of enzymes inside micrometer-sized liposomes. Life (Basel, Switzerland) **5**(1), 969–996 (2015). https://doi.org/10.3390/life5010969
6. Damiano, L., Canamero, L.: The frontier of synthetic knowledge: toward a constructivist science. World Futures **68**(3), 171–177 (2012). https://doi.org/10.1080/02604027.2012.668409
7. Elani, Y.: Construction of membrane-bound artificial cells using microfluidics: a new frontier in bottom-up synthetic biology. Biochem. Soc. Trans. **44**(3), 723–730 (2016). https://doi.org/10.1042/BST20160052
8. Exterkate, M., Caforio, A., Stuart, M.C.A., Driessen, A.J.M.: Growing membranes in vitro by continuous phospholipid biosynthesis from free fatty acids. ACS Synth. Biol. **7**(1), 153–165 (2018). https://doi.org/10.1021/acssynbio.7b00265
9. Fanti, A., Gammuto, L., Mavelli, F., Stano, P., Marangoni, R.: Do protocells preferentially retain macromolecular solutes upon division/fragmentation? A study based on the extrusion of POPC giant vesicles. Integr. Biol. (Camb) **10**(1), 6–17 (2018). https://doi.org/10.1039/c7ib00138j
10. Fox, S.V.: Self-assembly of the protocell from a self-ordered polymer. J. Sci. Ind. Res. **27**, 267–274 (1968)
11. Gebicki, J.M., Hicks, M.: Ufasomes are stable particles surrounded by unsaturated fatty acid membranes. Nature **243**(5404), 232–234 (1973)
12. Hadorn, M., Boenzli, E., Sørensen, K.T., De Lucrezia, D., Hanczyc, M.M., Yomo, T.: Defined DNA-mediated assemblies of gene-expressing giant unilamellar vesicles. Langmuir **29**(49), 15309–15319 (2013). https://doi.org/10.1021/la402621r
13. Hanczyc, M.M.: The early history of protocells - the search for the recipe of life. In: Rasmussen, S., et al. (eds.) Protocells Bridging Nonliving and Living Matter. MIT Press, Cambridge (2009)
14. Hanczyc, M.M., Fujikawa, S.M., Szostak, J.W.: Experimental models of primitive cellular compartments: encapsulation, growth, and division. Science **302**(5645), 618–622 (2003). https://doi.org/10.1126/science.1089904
15. Hargreaves, W.R., Deamer, D.W.: Liposomes from ionic, single-chain amphiphiles. Biochemistry **17**(18), 3759–3768 (1978)
16. Hosoda, K., Sunami, T., Kazuta, Y., Matsuura, T., Suzuki, H., Yomo, T.: Quantitative study of the structure of multilamellar giant liposomes as a container of protein synthesis reaction. Langmuir **24**(23), 13540–13548 (2008). https://doi.org/10.1021/la802432f

17. Kaneko, K.: Life: An Introduction to Complex Systems Biology. UCS. Springer, Heidelberg (2006). https://doi.org/10.1007/978-3-540-32667-0

18. Kita, H., et al.: Replication of genetic information with self-encoded replicase in liposomes. ChemBioChem. **9**(15), 2403–2410 (2008). https://doi.org/10.1002/cbic.200800360

19. Kuruma, Y., Stano, P., Ueda, T., Luisi, P.L.: A synthetic biology approach to the construction of membrane proteins in semi-synthetic minimal cells. Biochim. Biophys. Acta **1788**(2), 567–574 (2009). https://doi.org/10.1016/j.bbamem.2008.10.017

20. Lazzerini-Ospri, L., Stano, P., Luisi, P., Marangoni, R.: Characterization of the emergent properties of a synthetic quasi-cellular system. BMC Bioinform. **13**(Suppl 4), S9 (2012). https://doi.org/10.1186/1471-2105-13-S4-S9

21. Luisi, P.L.: Autopoiesis: a review and a reappraisal. Naturwissenschaften **90**(2), 49–59 (2003). https://doi.org/10.1007/s00114-002-0389-9

22. Luisi, P.L., Allegretti, M., Pereira de Souza, T., Steiniger, F., Fahr, A., Stano, P.: Spontaneous protein crowding in liposomes: a new vista for the origin of cellular metabolism. ChemBioChem **11**(14), 1989–1992 (2010). https://doi.org/10.1002/cbic.201000381

23. Maturana, H.R., Varela, F.J.: Autopoiesis and Cognition: The Realization of the Living. D. Reidel Publishing Company, 1st edn. (1980)

24. Mavelli, F.: Stochastic simulations of minimal cells: the Ribocell model. BMC Bioinform. **13**(Suppl 4), S10 (2012). https://doi.org/10.1186/1471-2105-13-S4-S10

25. Mavelli, F., Stano, P.: Experiments on and numerical modeling of the capture and concentration of transcription-translation machinery inside vesicles. Artif. Life **21**(4), 445–463 (2015)

26. Nakano, T., Eckford, A.W., Haraguchi, T.: Molecular Communications. Cambridge University Press, Cambridge UK (2013)

27. Oparin, A.I.: The pathways of the primary development of metabolism and artificial modeling of this development in coacervate drops. In: The Origins of Prebiological Systems and of their Molecular Matrices, pp. 331–345. S. W. Fox, New York, Academic Press edn. (1965)

28. Paradisi, P., Allegrini, P., Chiarugi, D.: A renewal model for the emergence of anomalous solute crowding in liposomes. BMC Syst. Biol. **9**(Suppl 3), S7 (2015). https://doi.org/10.1186/1752-0509-9-S3-S7

29. Rampioni, G., et al.: Synthetic cells produce a quorum sensing chemical signal perceived by Pseudomonas aeruginosa. Chem. Commun. **54**, 2090–2093 (2018). https://doi.org/10.1039/C7CC09678J

30. Schmidli, P., Schurtenberger, P., Luisi, P.: Liposome-mediated enzymatic-synthesis of phosphatidylcholine as an approach to self-replicating liposomes. J. Am. Chem. Soc. **113**(21), 8127–8130 (1991). https://doi.org/10.1021/ja00021a043

31. Pereira de Souza, T., Stano, P., Luisi, P.L.: The minimal size of liposome-based model cells brings about a remarkably enhanced entrapment and protein synthesis. Chembiochem **10**(6), 1056–1063 (2009). https://doi.org/10.1002/cbic.200800810

32. de Souza, T.P., Fahr, A., Luisi, P.L., Stano, P.: Spontaneous encapsulation and concentration of biological macromolecules in liposomes: an intriguing phenomenon and its relevance in origins of life. J. Mol. Evol. **79**(5–6), 179–192 (2014). https://doi.org/10.1007/s00239-014-9655-7

33. Souza, T.P.D., et al.: Vesicle aggregates as a model for primitive cellular assemblies. Phys. Chem. Chem. Phys. **19**(30), 20082–20092 (2017). https://doi.org/10.1039/C7CP03751A

34. Stano, P., Wehrli, E., Luisi, P.L.: Insights into the self-reproduction of oleate vesicles. J. Phys. Condens. Matter **18**(33), S2231 (2006). https://doi.org/10.1088/0953-8984/18/33/S37

35. Stano, P., Carrara, P., Kuruma, Y., de Souza, T.P., Luisi, P.L.: Compartmentalized reactions as a case of soft-matter biotechnology: synthesis of proteins and nucleic acids inside lipid vesicles. J. Mater. Chem. **21**(47), 18887–18902 (2011). https://doi.org/10.1039/c1jm12298c

36. Stano, P., D'Aguanno, E., Bolz, J., Fahr, A., Luisi, P.L.: A remarkable self-organization process as the origin of primitive functional cells. Angew. Chemie-Int. Ed. **52**(50), 13397–13400 (2013). https://doi.org/10.1002/anie.201306613

37. Stano, P., Luisi, P.: Theory and construction of semi-synthetic minimal cells. In: Nesbeth, D.N. (ed.) Synthetic Biology Handbook, pp. 209–258. CRC Press, London (2016)

38. Stano, P., Mavelli, F.: Protocells models in origin of life and synthetic biology. Life **5**(4), 1700–1702 (2015). https://doi.org/10.3390/life5041700

39. Stano, P., et al.: Recent biophysical issues about the preparation of solute-filled lipid vesicles. Mech. Adv. Mater. Struct. **22**(9), 748–759 (2015). https://doi.org/10.1080/15376494.2013.857743

40. van Swaay, D., deMello, A.: Microfluidic methods for forming liposomes. Lab. Chip. **13**(5), 752–767 (2013). https://doi.org/10.1039/c2lc41121k

41. Walde, P., Wick, R., Fresta, M., Mangone, A., Luisi, P.: Autopoietic self-reproduction of fatty-acid vesicles. J. Am. Chem. Soc. **116**(26), 11649–11654 (1994). https://doi.org/10.1039/c2lc41121k

42. Weiss, M., et al.: Sequential bottom-up assembly of mechanically stabilized synthetic cells by microfluidics. Nat. Mater. **17**(1), 89–96 (2018). https://doi.org/10.1038/nmat5005

43. Zepik, H.H., Blochliger, E., Luisi, P.L.: A chemical model of homeostasis. Angew. Chem.-Int. Edit. **40**(1), 199–202 (2001)

44. Zhu, T.F., Szostak, J.W.: Coupled growth and division of model protocell membranes. J. Am. Chem. Soc. **131**(15), 5705–5713 (2009). https://doi.org/10.1021/ja900919c

Towards the Synthesis of Photo-Autotrophic Protocells

Emiliano Altamura[1], Paola Albanese[1], Roberto Marotta[2],
Pasquale Stano[3], Francesco Milano[4], Massimo Trotta[4],
and Fabio Mavelli[1,5(✉)]

[1] Chemistry Department, University "Aldo Moro",
Via Orabona 4, 70126 Bari, Italy
fabio.mavelli@uniba.it
[2] Electron Microscopy Facility, Istituto Italiano di Tecnologia, IIT,
Via Morego 30, 16163 Genoa, Italy
[3] Department of Biological and Environmental Sciences and Technologies
(DiSTeBA), University of Salento, Ecotekne, 73100 Lecce, Italy
[4] CNR-IPCF, Istituto per i Processi Chimico Fisici,
Via Orabona 4, 70126 Bari, Italy
[5] CNR-NANOTECH, Istituto di Nanotecnologia,
Via Orabona 4, 70126 Bari, Italy

Abstract. In this contribution we discuss the possible strategies to synthesize photo-autotrophic artificial protocells starting from scratch, following the semi-synthetic bottom up approach. The main aim is to build up artificial compartmentalized systems able to mimic living cell behavior in the transduction of light energy in chemical energy. Some preliminary results and future perspective are presented and discussed.

Keywords: Artificial photosynthesis · Membrane protein ·
Photo-chemical oscillations · Droplet transfer method · Giant vesicles ·
Light transduction · Photosynthetic reaction center · bc1 · Synthetic cell

1 Introduction

1.1 ASAPs: Artificial Simplified Autotrophic Protocells

The synthesis of artificial cells starting from inanimate artificial or natural compounds is an ambitious scientific goal [1–6] that was postulated initially as an issue in the origin-of-life researches [7–10], but it has fast gained attention in the past few years. In fact, the rapid expansion of synthetic biology [11] has given additional conceptual stimuli and technical tools to this field, especially by the so-called bottom-up approach [12]. In this framework, it has been shown that vesicle-based cell-like systems (shortly "protocells", a term often used to indicate both primordial cells or simplified cellular prototypes) can be designed and assembled to perform specific function (for biotechnological applications) and for studies in the origin-of-life field.

We recently focused our attention on the construction of synthetic cells capable to convert light energy into chemical energy in form of proton gradient across the vesicle

© Springer Nature Switzerland AG 2019
M. Bartoletti et al. (Eds.): CIBB 2017, LNBI 10834, pp. 186–199, 2019.
https://doi.org/10.1007/978-3-030-14160-8_18

membrane [13, 14] that can eventually be exploited to synthetize high free energy compound, like ATP and/or NADH. To achieve this aim we have been inspired by natural systems trying to mimic the light phase of the photosynthesis performed by bacteria [15]. Two main different strategies are followed: the single and the multi compartment approaches (SCA and MCA respectively). In SCA, we try to reconstitute in the lipid membrane of giant unilamellar vesicles (GUVs) all the protein complexes involved in the light phase of bacterial photosynthesis: the reaction center (RC), the coenzyme Q–cytochrome c oxidoreductase (bc1), and the ATP synthase (ATP-syn). GUVs are spherical aqueous compartments closed by a lipid double layer with diameter in the range of tenth of micrometers, that are self-aggregate artificial structures suitable to mimic the cellular morphology. On the other hand, in MCA, instead of extracting each single photosynthetic enzyme from bacteria and reconstituting all of them in the vesicle membrane, we optimize a procedure for extracting cromatophores, small natural organelles (radius 20–50 nm), that contain all the photosynthetic apparatus in their membrane. The cromatophores can be then entrapped in the internal aqueous lumen of GUVs, in order to implement multi-compartment systems able to transduce light energy.

Therefore, in both approaches, the final goal is to prepare photo-autotrophic synthetic protocells able to convert ADP into ATP molecules driven by light. In this paper we describe the steps already done to achieve this ambitious goal following both the mentioned approaches and the further moves to be accomplished in a close future. Before doing this, in the next section we briefly describe the photosynthetic apparatus present in bacteria.

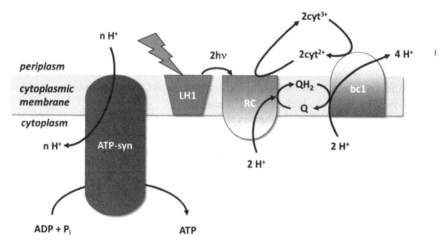

Fig. 1. Schematic representation of the ternary RC/bc1/ATP-syn system in its physiological orientation. After visible irradiation, the antenna complex LH1 absorbs and transfers the light energy to bacteriochlorophylls dimer of RC that is photo-oxidized. 2 electrons are then translocated to a ubiquinone molecule (Q), which is reduced to ubiquinol QH_2 with protons taken up from the inner membrane side (cytosol side). The photo-oxidized dimer is reduced back to its initial state by the reduced cytochrome c2 (cyt^{2+}), which in turn becomes oxidized (cyt^{3+}). The two newly produced species (QH_2 and cyt^{3+}) are converted back to Q and cyt^{2+} by the coenzyme Q–cytochrome c oxidoreductase (bc1). A proton gradient (alkaline inside) is thus obtained. Next, ATP synthase (ATP-syn) exploits the proton gradient to produce ATP from ADP and inorganic phosphate Pi in the internal side of the membrane.

1.2   The Bacterial Photosynthetic Apparatus

Photosynthesis is responsible for the photochemical conversion of light into the chemical energy that fuels the planet Earth. The protein complexes that are involved in this process are schematically reproduced in Fig. 1 and they are located in the cytoplasmic membrane (CM) of photosynthetic bacteria [15]. The photochemical core of this process in all photosynthetic organisms is the transmembrane protein called the Reaction Center (RC) a transmembrane protein complex composed of one mostly hydrophilic H subunit and two highly hydrophobic subunits, L and M. In purple photosynthetic bacteria, this photo-enzyme is located in CM surrounded by antenna complexes LH1 and LH2 that absorb light energy and transfer it to reaction centers. In fact, when the energy of a photon is transferred to RC, the photo-cycle starts and an electron-hole pair is generated. In the presence of the reduced cytochrome cyt^{2+} (an electron donor) in the external pool, electrons are transferred from cyt^{2+} to RC while protons are taken up from the cytoplasm by ubiquinone Q (an electron acceptor) present in an internal site located in the H hydrophilic subunit. This process produces a molecule of ubiquinol QH_2 that eventually leaves its binding site. The stoichiometry of this first step is as follows:

$$2cyt^{2+} + Q + 2H_{in}^+ \xrightarrow{RC,h} 2cyt^{3+} + QH_2 \tag{1}$$

The photocycle is then completed by the activity of the coenzyme Q–cytochrome c oxidoreductase (bc1) that catalyses the opposite reaction without consuming light energy:

$$2cyt^{3+} + QH_2 + 2H_{in}^+ \xrightarrow{bc1} 2cyt^{2+} + Q + 4H_{out}^+ \tag{2}$$

The result of the photo-cycle, Eqs. (1) and (2), is the net translocation of 4 protons across the cytoplasmic membrane and the formation of a transmembrane proton gradient driven by the transduction of light energy.

This proton motive force is then exploited by the ATP synthase, an enzyme able to convert ADP in ATP molecules in presence of endogenous phosphate groups Pi in the cytoplasm. In this way, photo-autotrophic cells fuel the whole metabolism of the organism.

It is worthwhile to stress that the physiological orientation of the three enzymatic complexes must be preserved when transmembrane proteins are reconstituted in the artificial protocell membrane, in order to make the photosynthetic apparatus perfectly functioning. This is, of course, a crucial requirement in the optimization of the preparation procedure of ASAPs. To overcome this drawback, an alternative strategy is to extract from photosynthetic bacteria nano-sized vesicles: the chromatophores, containing all the photosynthetic apparatus retaining the physiological orientation [15, 16]. A schematic draw of a chromatophore is reported in Fig. 2 [17]. Chromatophores are usually originated from invaginations of the cytoplasmic membrane, mostly produced when bacteria grow photosynthetically in anaerobic conditions. They can be collected by breaking the CM membrane with a suitable procedure and separating them from others cellular fragments.

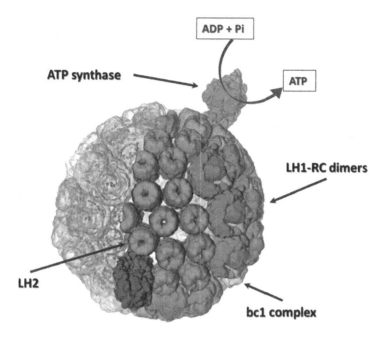

Fig. 2. Atomic structural model of a chromatophore vesicle from R. sphaeroides, the inner radius is 50 nm. Lipid molecules among protein complexes are not shown for sake of simplicity. Reproduced with modifications from [17] according to CC by 4.0 license.

2   Materials and Methods

In this section some experimental procedures common to both SCA and MCA will be briefly described.

2.1   Giant Unilamellar Vesicles Preparation: The Phase Transfer Method

Giant unilamellar vesicles (GUVs) are spherical artificial compartments with a diameter in the range of tenths of micrometers enveloped by a bilayer of lipids. For their size and morphology, they represent suitable *in vitro* models for real cells. GUVs can be prepared following different protocols [18]: natural swelling and electro formation are the more traditional ones, but recently the lipid-coated ice droplet hydration method [19], the phase transfer method [20] have been also presented and microfluidic apparatus developed [21]. In this paper we used the droplet transfer method [20] since this method does not require a complex instrumentation or manipulation procedure. Moreover, it needs a small amount of chemicals and exhibits high rate of success and high yield of solute encapsulation. This method starts by preparing a water in oil macroemulsion [13, 20]. The water droplets, swimming in the organic phase, are enveloped by just a single monolayer of phospholipids, and they represent the precursors of GUVs. The macroemulsion is obtained simply pipetting by hand a water solution (the Inner Solution IS) in the organic phase composed of mineral oil in presence of a single kind of phospholipid: 1-palmitoyl-2-oleoyl-*sn*-glycero-3-phosphocholine (POPC). IS must contain all the compounds that have to be entrapped in the internal volume of the GUVs and these molecules remain confined in the core of the droplets. In order to add the second layer to the droplets to form the bilayered membrane of GUVs, the macroemulsion is stratified in a 1,5 mL Eppendorf tube on a water phase (the Outer Solution OS) on top of which it has been previously stratified a mineral oil layer rich with POPC. By centrifugation, the droplets are then forced to go from the upper organic phase to the lower water phase and, by crossing the POPC-rich interphase, they take up the second layer. Finally, GUVs can be collected as a pellet at the bottom of the test tube after centrifugation, thanks to a density difference between the IS and OS obtained by adding sugars with different molecular weights, keeping isotonic conditions. The pellet, then, can be washed and re-suspended in an isotonic aqueous solution obtaining a vesicle suspension with the external and internal water phase of the desired compositions. In Fig. 3, a schematic draw of this procedure is reported in the left panel (a), while a picture of the pellet of GUVs obtained after centrifugation is illustrated in the middle panel (b). For further details the readers are recommended to refer to the original papers [13, 20].

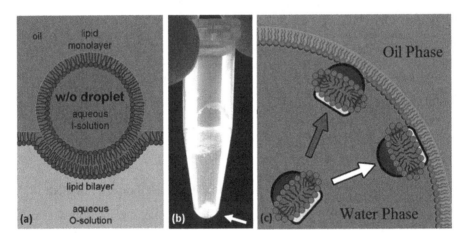

Fig. 3. The phase transfer method: (a) schematic draw of the formation of the second layer during the droplets centrifugation; (b) the pellet obtained at the bottom of the test tube pointed out by the arrow; (c) a possible mechanism of the encapsulation of membrane proteins extracted from living systems in form of micellar aggregates and added to the inner solution. The white arrow indicates the favored encapsulation of a RC@micelle aggregate driven by the hydrophilic subunit. Reproduced from [13] with permission.

2.2   Chromatophores Extraction from *Rhodobacter sphaeroides*

Cells of *Rh. sphaeroides* strain R26 were purchased from the Deutsche Sammlung von Mikroorganismen und Zellkulturen GmbH (DSMZ, catalog number 2340). Bacterial cells were grown, first, oxygenically in the dark for 12 h, then photoheterotrophically under anaerobic conditions at pH 6.9 illuminating the bottles with a 100 W tungsten filament light bulb placed at 25 cm from the vessels. Harvested cells were washed twice and re-suspended in a 5 mM K-Phosphate buffer (Ph = 8). A few flakes of DNase and 5 mM $MgSO_4$ were added. After 15 min incubation, the cells were disrupted by a single French press step operating at 15 MPa and 4 °C. In these conditions cell disruption promotes the closure of the CM invagination to form sealed vesicles with diameter of roughly 50–100 nm bearing the intact photosynthetic apparatus: RC, LHI, LHII, bc1 complex, ATP synthase and the water-soluble cytochrome cyt^{2+} in the inner aqueous core. The sample was centrifuged at 4 °C, for 15 min obtaining a pellet containing the cell debris and a supernatant containing the chromatophores. The supernatant solution was then ultra-centrifuged at 140000 g for 120 min and the obtained pellet was suspended in the same buffer containing chromatophores roughly at a $OD_{860} = 50$.

2.3   Proteins Extraction from *Rhodobacter sphaeroides*

Extraction and Purification of the Reaction Center (RC). RC was extracted from blue-green strain R-26 of the photosynthetic bacterium *Rhodobacter sphaeroides* following an established procedure using lauryldimethylamine N-oxide (LDAO), then it has been precipitated with ammonium sulphate and purified with a diethylaminoethyl (DEAE) Sephadex G25 chromatography [22]. The final ratio of absorbances at 280 nm and 802 nm (A280/A802) was lower than 1.3 (the optimal value for pure RC is 1.28). The RC concentration was determined spectrophotometrically measuring the absorption at 802 nm of the steady-state using 288,000 M^{-1} cm^{-1} as extinction coefficient [23]. Usually the stock solution of RC in LDAO 0.025% is obtained at the final concentration of around 70 μM.

Extraction and Purification of the Coenzyme Q–Cytochrome C Oxidoreductase (bc1). Extraction and purification of bc1 were performed as previously described in literature [24]. One volume of chromatophores suspension ($OD_{860} = 50$) was diluted with 0.3 volumes of buffer (50 mM MOPS (pH 7.8), 100 mM NaCl, 1 mM $MgSO_4$, 20% (w/v) glycerol, and 5 mM histidine) and 0.7 volumes of Ni-NTA resin equilibrated in buffer B. Dodecyl maltoside (DM) from Sigma was added dropwise from a stock solution 10% (w/v) to a final concentration of 0.6% (w/v). After 40 min incubation at 4 °C, the mixture was transferred to a chromatographic column and washed with about 20 column volumes of the buffer until the eluent was clear. Therefore, the purified bc1 was eluted with a solution containing 200 mM histidine and 15 μg/mL POPC (Avanti Polar-Lipids, Inc., Alabaster, AL). The histidine was removed by overnight dialysis against 50 mM MOPS (pH 7.0), 100 mM NaCl, 1 mM $MgSO_4$, 20% (w/v) glycerol, 0.01% DM, and 15 μg/mL POPC.

2.4   CrioTEM Analysis of Cromatophores

The membrane fragment suspension derived by cromatophores extraction procedure were vitrified by applying a 3 μL aliquot to a previously glow-discharged 200-mesh 2/1 Quantifoil carbon grids (Ted Pella, USA). Grids were blotted and then plunged into liquid ethane using a FEI Vitrobot Mark IV (FEI Company, the Netherlands). The samples were imaged in bright field transmission electron microscopy (TEM) using a

Tecnai G2F20 microscope (FEI Company, the Netherlands) operating at an acceleration voltage of 200 kV and equipped with a Schottky Field Emission electron source and a US1000 2kx2k Gatan charge-coupled device (CCD) camera. The cryo-EM imaging was performed under low dose conditions (with a total dose of 60–80 electrons/Å^2).

2.5   ATP Production by Chemiluminescent Assay

In order to monitor the conversion of ADP in ATP, the light produced by the ATP-dependent oxidative carboxylation of luciferin by firefly luciferase enzyme has been followed. In fact, this enzymatic reaction takes place only in presence of ATP and consists in the oxidation of the luciferin in oxyluciferin, a compound in an excited state that relaxes emitting energy in form of light at wavelength 555 nm.

For this chemiluminescent assay, we prepared 10 mL of an enzymatic kit as follows: 8.9 mL H_2O, 0.5 mL 20X solution buffer (containing 500 mM Tricine buffer pH 7.8, 100 mM $MgSO_4$, 2 mM EDTA and 2 mM sodium azide), 0.1 mL 0.1 M DTT, 0.5 mL of 10 mM D-luciferin and 2.5 µL of firefly luciferase 5 mg/mL stock solution.

3   Results and Discussion

Obtained results and new data for both SCA and MCA will be presented and further steps and optimizations discussed in order to prepare really functioning ASAPs.

3.1   SCA: Single Compartment Approach

Reconstituting RC in GUVs: RC@GUV. In a previous paper [13], we have already shown as the phase transfer method is a suitable method for the reconstitution of transmembrane proteins in the GUV membrane. If the transmembrane proteins are extracted from living systems in form of a micellar suspension, this suspension can be added to the IS water solution before preparing the macroemulsion. Since transmembrane proteins have highly hydrophobic subunits (for instance the L and M subunits in the RC structure), these force the spontaneous embedding of the proteins in the vesicle membrane [13, 20]. In the case of the RC, since the H subunit is instead hydrophilic, it drives the embedding process to occur with a preferential orientation, determining a high percentage of uniform alignment of RCs, around 90% [13]. This orientation corresponds to the physiological one observed in the cytoplasmic membrane. This process is schematically represented in Fig. 3(c). It is worthwhile to note that in this configuration when photon energy is transferred to RCs the reaction reported in Eq. (1) occurs and protons are taken up from the internal aqueous vesicle lumen increasing the internal pH of GUVs.

Testing the RC@GUV Photo-Activity. The RC@GUV photo-activity can be proved by encapsulating in the vesicle internal lumen a fluorescent probe sensible to pH, like pyranine that increases its fluorescence emission as the pH increases. Therefore, RC@GUVs entrapping pyranine in the internal aqueous lumen RC@GUVs{pyranine} can be irradiated at 860 nm wavelength in order to promote the reaction described by Eq. (1). This produces an increase of the internal pH due to consumption of two protons necessary for the conversion of Q in QH_2 and it causes an increase of the encapsulated pyranine fluorescence, recorded at 520 nm. This experiment has been performed directly under a confocal microscope [13] and the result is shown in Fig. 4, while the experimental conditions are reported in Table 1.

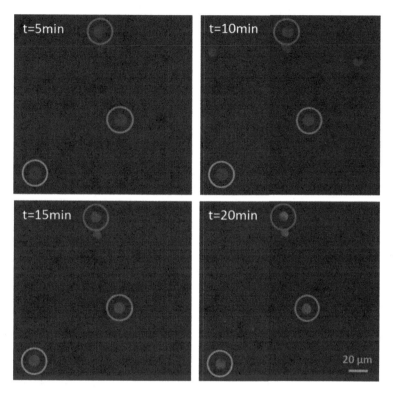

Fig. 4. Confocal microscopy images of RC@GUVs{pyranine} transducing light energy in a transmembrane pH gradient. The process is monitored by the increase of the internal fluorescence of the pyranine. The vesicle suspension has been irradiated at 860 nm directly under the microscope for 4 times intervals of 5 min and then each picture has been taken. The rate of pH increase has been determined equal to 0.061 ± 0.004 pH units per min.

Since in RC@GUVs{pyranine} RC works alone, *i.e.* it is not coupled with the bc1, the photo-cycle does not take place completely. Therefore, in order to restore the electron donor (the reduced cytochrome cyt^{2+}), $K_3Fe(CN)_6$ has been added to the external solution since this compound can spontaneously reduce cyt^{3+} according to the following reaction:

$$cyt^{3+} + Fe(CN)_6^{3-} \rightarrow cyt^{2+} + Fe(CN)_6^{4-} \tag{3}$$

Table 1. Overall concentrations of different compounds in the vesicle and micelle suspensions. For the complete set of the experimental conditions the reader must refer to the original papers.

Compound	Concentration giant vesicle Ref. [13]	Concentration micelles Ref. [26]
[RC]		30 nM
[bc1]	0.0	150 nM
[cyt^{2+}]	0.0	20.0 μM
[cyt^{3+}]	5.0 μM	0.0 μM
[Q]	60.0 μM	50.0 μM
[QH$_2$]	0.0	150 μM
[K$_3$Fe(CN)$_6$]	10.0 mM	0.0

Coupling RC and bc1. In order to test if the bc1 extracted from living bacteria is still active, it can be coupled with RC, by mixing the protein complex micellar suspensions: RC@micelles and bc1@micelles, in a test tube in presence of the needed substrates [26]. Moreover, this allows us to check if it is possible to implement in a simplified artificial system the photo-cycle occurring in the bacteria cytoplasmic membrane, Eqs. (1) and (2). In the case of a micellar suspension, it is important to stress that the light energy cannot be converted into a pH gradient, since a lipid membrane is not present. Therefore, the absorbed photon energy is dissipated in form of thermal energy, while the pH of the solution remains constant since all the protons are exchanged with the same water solution. The time course of the reaction can then be monitored by following the absorbance at 550 nm of the pairs cyt^{2+}/cyt^{3+} by exploiting the difference between cyt^{2+} and cyt^{3+} molar extinction coefficients (namely, $\varepsilon_{cyt2+} = 28$ mM^{-1}cm^{-1} and $\varepsilon_{cyt3+} = 9$ mM^{-1}cm^{-1}). Figure 5 shows repetitive light-driven oscillations, in highly reproducible way for 45 min. These A$_{550}$ oscillations are due to the oxidation/reduction of the couple cyt^{2+}/cyt^{3+} since the extinction coefficient ε_{cyt2+} is greater than ε_{cyt3+}. In these experimental conditions, irradiation drives the RC-catalysed reaction very efficiently, and cyt^{2+} gets completely oxidised and A$_{550}$ decreases from ca. 0.56 to ca. 0.20. Consequently, its concentration reaches a minimal stationary value that is balanced by the opposite – and simultaneously occurring – bc1-catalysed cyt^{3+} reduction. When the light is switched off, the RC activity is stopped and bc1 can restore the cyt^{2+} concentration almost to its initial value.

This experiment proves that RC and bc1 are both active in the investigated experimental conditions and the photo-cycle can take place in the artificial micellar system as expected. Moreover, Fig. 5 clearly shows that the light-driven RC-catalysed reaction is more efficient than bc1-catalysed one, in fact, when at the end of the light phase the stationary state is reached, the cyt^{2+} is almost completely consumed.

Optimizing the Photo-Cycle in GUVs. The further steps in SCA will be focused on the implementation of protocells: RC/bc1@GUV, and on the optimization of the enzymatic levels in order to increase the RC turnover and the efficiency of the light transduction. The reconstitution of bc1 in GUVs can be achieved by adding the bc1@micelles suspension in the inner solution like in the case of RC@micelles, or alternatively adding bc1@micelles at a solution of preformed RC@GUVs [25].

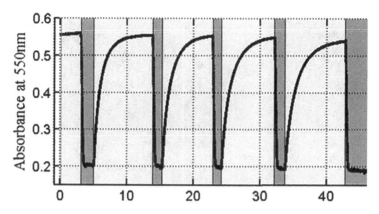

Fig. 5. Time course of the absorbance at 550 nm of a mixed RC@micelles and bc1@micelles solution ([RC] = 30 nM and [bc1] = 210 nm) under continuous light irradiation and in the dark (red and grey background) in presence of [cyt^{2+}] = 20 μM, [Q] = 50 μM and [QH$_2$] = 150 μM. Reproduced with permission from [26] (Color figure online)

This because in the case of bc1, its hydrophilic subunit is located in the cytoplasm, i.e. from the opposite side of the cellular membrane respect to the H hydrophilic subunit of RC (see Fig. 2). Regarding the optimization of the RC turnover, this can be done using the micellar systems as a prototype model for the implementation of the photo-cycle. What can be done is to increase the level of the bc1, more than the ration 1/7 of RC/bc1, as reported in Fig. 5. This should allow to increase the efficiency of the cyt^{3+} reduction and to rise the stationary rate at which both enzymes work under continuous illumination. The reconstitution in GUVs of bc1 extracted from mitochondria of mammalian cells could be another possible strategy. In fact, these enzymes, in spite of the similarity of the active site with those of the bacterial version, exhibit a higher enzymatic activity and a higher stability [27]. This will bring us to realize engineered hybrid protocells able to maximize the light energy transduction in form of a proton motive force, as a first milestone towards the synthesis of ASAPs following the Single Compartment Approach.

3.2 MCA: Multi Compartment Approach

In this section we describe some preliminary results obtained by following the MCA for the preparation of ASAPs. In particular, we have focused the attention in a procedure for extracting chromatophores from living bacteria and testing their activity in producing ATP under irradiation.

Morphological Analysis of Chromatophores. A procedure for the extraction of cromatophores from living bacteria has been optimized and described in the Materials and Methods section. The solution containing the cromatophores has been then analyzed by Cryo-TEM microscopy in order to check if the cytoplasmic membrane closes to form spherical nano-sized vesicles after the rupture in the French press (Fig. 6).

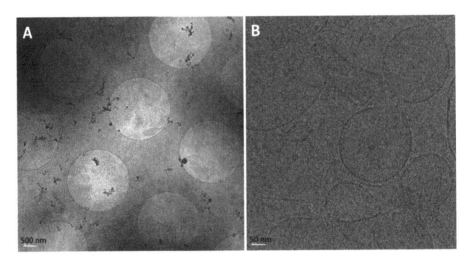

Fig. 6. Cryo transmission electron microscopy (Cryo-TEM) analysis. (A) Low magnification image of vitrified vesicles acquired at low electron dose. (B) Higher magnification showing vesicles of various size and shape

Cryo-TEM analysis shows that aggregates of different morphology are present in the obtained suspension, most of all are closed vesicles with an average diameter $D = 116 \pm 69$ nm (average done on 76 collected items), but some of them exhibit elongate forms and some open membrane fragments are also presents, as it is shown in Fig. 6.

Testing the Chromatophores Activity. To check if the extraction procedure brings to cromatophores that keep the activity of all the photosynthetic protein complexes we checked if ADP added to the bacterial extract suspension can be converted in ATP molecules by irradiating the solution. To reveal the presence of ATP a well-known chemiluminescent assay, the enzymatic oxidation of luciferin, has been used. The plot in Fig. 7 shows the time course of the 555 nm chemiluminescence registered spectrophotometrically from a cuvette filled with the enzymatic kit, after the addition of ADP and cromatophores, and after switching on/off the light irradiation at 860 nm. The enzymatic kit alone exhibits a negligible chemiluminesence that undergoes a small jump when ADP is added to the kit, probably due to some ATP present as impurity. After the chromatophores addition the light emission increases linearly and this can be due to a residual activity of cromatophores driven by the pH gradient that occurs across the organelles and the buffered solution of the enzymatic kit. When the cuvette is irradiated at the wavelength 860 nm corresponding to the RC excitation, the signal increases abruptly, proving that the cromatophores start to produce ATP and this speed up the firefly luciferin enzymatic activity. When the irradiation is switched off, what is observed is a decrease of the signal due to the natural decay of the excited oxyluciferin. If the light is switched on again, the emission increases with the same rate as before, reaching quite soon the instrumental saturation of the signal. Switching off the irradiation results in a new decay that becomes visible after almost one minute, time needed for the consumption of the excess of the exited product. This experiment proves that the fraction of integer chromatophores keep their physiological activity and they can produce ATP driven by light in the external solution.

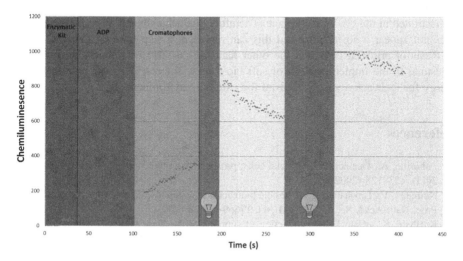

Fig. 7. Chemiluminesence time course monitored at 555 nm to prove the activity of chromatophores. The enzymatic kit alone (light yellow time window) does not exhibit bioluminescence. After the addition of ADP (dark yellow) a small increase in the signal is obtained, maybe due to ATP present as contamination. Adding the cromatophores suspension in the cuvette (dark green) the enzymatic reaction takes place at slow rate. When the light is switched on (first red sector) the enzymatic reaction speeds up. Switching off the light (gray), the reaction stops and the decay of the product reduce the emission. If the light is switched on again, the emission restarts at the same rate as before and the signal rapidly reaches a saturation condition. By switching off the light, it takes almost a minute before observing a new decay in chemiluminesence, since the accumulated product has to be consumed. (Color figure online)

Towards the Preparation of ASAPs. The next move towards the preparation of ASAPs is to optimize the encapsulation of the chromatophores extracted by bacteria in GUVs and proves that in this multi-compartment systems ATP can be produced driven by light irradiation. For this purpose, the chemiluminesence assay cannot be suitable due to the large light scattering of the GUVs suspension. Therefore, an alternative strategy can be the use of a probe that can become fluorescent in presence of the ATP molecules [28]. In fact, this will allow to preformed activity test directly by the confocal microscope. Moreover, a procedure for the purification of the chromatophores from cytoplasmic open membrane fragments can also optimize the behavior of ASAPs.

4 Conclusion

In this contribution two possible strategies finalized to the synthesis of autotrophic simplified artificial protocells (ASAPs) have been discussed recalling some recent results and presenting new experimental data. Although, with the Single Compartment Approach it has been possible to prepare protocells able to transduce light energy in a pH transmembrane gradient, however, the Multi Compartment Approach seems more promising, since it avoids the problem of a correct protein alignment when enzymes are reconstituted in the GUVs membrane. Therefore, the achievement of preparing ASAPs does not appear a so far goal and this can represent a step forward the challenging synthesis of artificial cells. On the other hand, it can also be sejen as an important milestone in the implementation of soft bio-robots, since ASAPs are able to get the energy for their activity directly from the external environment.

References

1. Pohorille, A., Deamer, D.: Artificial cells: prospects for biotechnology. Trends Biotechnol. **20**(3), 123–128 (2002)
2. Noireaux, V., Libchaber, A.: A vesicle bioreactor as a step toward an artificial cell assembly. Proc. Natl. Acad. Sci. U.S.A. **101**(51), 17669–17674 (2004)
3. Kurihara, K., et al.: Self-reproduction of supramolecular giant vesicles combined with the amplification of encapsulated DNA. Nat. Chem. **3**(10), 775–781 (2011)
4. Kuruma, Y., Stano, P., Ueda, T., Luisi, P.L.: A synthetic biology approach to the construction of membrane proteins in semi-synthetic minimal cells. Biochim. Biophys. Acta **1788**(2), 567–574 (2009)
5. Lentini, R., et al.: Integrating artificial with natural cells to translate chemical messages that direct E. coli behaviour. Nat. Commun. **5**, 4012 (2014)
6. Fujii, S., et al.: Liposome display for in vitro selection and evolution of membrane proteins. Nat. Protoc. **9**(7), 1578–1591 (2014)
7. Morowitz, H.J., Heinz, B., Deamer, D.W.: The chemical logic of a minimum protocell. Orig. Life Evol. Biosph. **18**(3), 281–287 (1988)
8. Oberholzer, T., Wick, R., Luisi, P.L., Biebricher, C.K.: Enzymatic RNA replication in selfreproducing vesicles: an approach to a minimal cell. Biochem. Biophys. Res. Commun. **207**(1), 250–257 (1995)
9. Szostak, J.W., Bartel, D.P., Luisi, P.L.: Synthesizing life. Nature **409**(6818), 387–390 (2001)
10. Mansy, S.S., Szostak, J.W.: Reconstructing the emergence of cellular life through the synthesis of model protocells. Cold Spring Harb. Symp. Quant. Biol. **74**, 47–54 (2009)
11. de Lorenzo, V., Danchin, A.: Synthetic biology: discovering new worlds and new words. EMBO Rep. **9**(9), 822–827 (2008)
12. Luisi, P.L., Ferri, F., Stano, P.: Approaches to semi-synthetic minimal cells: a review. Naturwissenschaften **93**(1), 1–13 (2006)
13. Altamura, E., et al.: Highly oriented photosynthetic reaction centers generate a proton gradient in synthetic protocells. PNAS **14**(15), 3837–3842 (2017)
14. Stano, P., Altamura, E., Mavelli, F.: Novel directions in molecular systems design: the case of light-transducing synthetic cells. Commun. Integr. Biol. **10**(1–7), e1365993 (2017)

15. Altamura, E., Mavelli, F., Milano, F., Trotta, M.: Photosynthesis without the organisms: the bacterial chromatophores. In: Piotto, S., Rossi, F., Concilio, S., Reverchon, E., Cattaneo, G. (eds.) Advances in Bionanomaterials, Lecture Notes in Bioengineering, pp. 165–175. Springer, Cham (2018)

16. Scheuring, S., et al.: The architecture of Rhodobacter sphaeroides chromatophores. Biochimica et Biophysica Acta **1837**, 1263–1270 (2014)

17. Sener, M., Strumpfer, J., Singharoy, A., Hunter, C.N., Schulten, K.: Overall energy conversion efficiency of a photosynthetic vesicle. eLife **5**, e09541 (2016)

18. Walde, P., Cosentino, K., Engel, H., Stano, P.: Giant vesicles: preparations and applications. ChemBioChem **11**, 848–865 (2010)

19. Sugiura, S., et al.: Novel method for obtaining homogeneous giant vesicles from a monodisperse water-in-oil emulsion prepared with a microfluidic device. Langmuir **24**(9), 4581–4588 (2008)

20. Pautot, S., Frisken, B.J., Weitz, D.A.: Engineering asymmetric vesicles. Proc. Natl. Acad. Sci. U.S.A. **100**, 10718–10721 (2003)

21. Matosevic, S., Paegel, B.M.: Stepwise synthesis of giant unilamellar vesicles on a microfluidic assembly line. J. Am. Chem. Soc. **133**(9), 2798–2800 (2011)

22. Milano, F., Italiano, F., Agostiano, A., Trotta, M.: Characterisation of RC-proteoliposomes at different RC/lipid ratios. Photosynth. Res. **100**, 107–112 (2009)

23. Straley, S.C., Parson, W.W., Mauzerall, D.C., Clayton, R.K.: Pigment content and molar extinction coefficients of photochemical reaction centers from Rhodopseudomonas spheroides. Biochim. Biophys. Acta **305**(3), 597–609 (1973)

24. Guergova-Kuras, M., Salcedo-Hernandez, R., Bechmann, G., Kuras, R., Gennis, R.B., Crofts, A.R.: Expression and one-step purification of a fully active polyhistidine-tagged cytochrome bc1 complex from Rhodobacter sphaeroides. Protein Expr. Purif. **15**, 370–380 (1999)

25. Yanagisawa, M., Iwamoto, M., Kato, A., Yoshikawa, K., Oiki, S.: Oriented reconstitution of a membrane protein in a giant unilamellar vesicle: experimental verification with the potassium channel KcsA. J. Am. Chem. Soc. **133**(30), 11774–11779 (2011)

26. Altamura, E., et al.: First moves towards photoautotrophic synthetic cells: in vitro study of photosynthetic reaction centre and cytochrome bc1 complex interactions. Biophys. Chem. **229**, 46–56 (2017)

27. Trumpower, B.L.: Cytochrome bc1 complex (Respiratory Chain Complex III). In: Encyclopedia of Biological Chemistry, vol. 1, pp. 528–534. Elsevier (2004)

28. Sunnapu, O., et al.: Rhodamine-based fluorescent turn-on probe for facile sensing and imaging of ATP in mitochondria. ChemistrySelect **2**, 7654–7658 (2017)

Hierarchical Block Matrix Approach for Multi-view Clustering

Angela Serra[1,2,3]([✉]) [iD], Maria Domenica Guida[1], Pietro Lió[4] [iD],
and Roberto Tagliaferri[1] [iD]

[1] NeuRoNe Lab, DISA-MIS, University of Salerno, Fisciano, SA, Italy
`aserra@unisa.it`
[2] Faculty of Medicine and Life Sciences, University of Tampere, Tampere, Finland
[3] Institute of Biosciences and Medical Technologies (BioMediTech),
Tampere, Finland
[4] Computer Laboratory, University of Cambridge, Cambridge, UK

Abstract. Scientists are facing two important challenges when investigating life processes. First, biological systems, from gene regulation to physiological mechanisms, are inherently multiscale. Second, complex disease data collection is an expensive process, and yet the analyses are presented in a rather empirical and sometimes simplistic way, completely missing the opportunity of uncovering patterns of predictive relationships and meaningful profiles. In this work, we propose a multi-view clustering methodology that, although quite general, could be used to identify patient subgroups, for different omic information, by studying the hierarchical structures of the patient data in each view and merging their topologies. We first demonstrate the ability of our method to identify hierarchical structures in synthetic data sets and then apply it to real multi-view multi-omic data sets. Our results, although preliminary, suggest that this methodology outperforms single-view clustering approaches and could open several directions for improvements.

Keywords: Multi-view clustering · Patient sub-typing ·
Hierarchical block matrices

1 Introduction

The technological advancement and the richness and variety of available data sets have opened new horizons for investigators in the bio-medical field. One of the last challenges is to create learning models based on heterogeneous data sources. Multi-view learning methodologies concerns with the analysis of multi-model data, where the patterns are represented by different sets of features extracted from multiple data sources [17]. Multi-view learning techniques are usually divided into early, intermediate or late integration, depending on the stage of the analysis when the integration is performed [17]. In early integration, data are concatenated to create a single feature space before starting the analysis.

© Springer Nature Switzerland AG 2019
M. Bartoletti et al. (Eds.): CIBB 2017, LNBI 10834, pp. 200–212, 2019.
https://doi.org/10.1007/978-3-030-14160-8_19

In intermediate integration, all the single views are translated into a common space (i.e. a kernel) and then combined in order to be analysed. Finally, in late integration, each view is analysed independently and then the results are linked together.

Multi-view learning has been applied also to improve classical clustering learning algorithms making them capable to identify data structures by taking into account different feature sets in order to provide a deeper understanding of the underlying principles governing complex systems. Different methods were proposed such as those based on matrix factorisation that integrate clustering solutions obtained on each single view [8,12,16]. Other approaches use modifications of the classical clustering algorithms such as k-means [2]. Other methods, instead, are based on canonical correlation analysis to perform multi-view clustering [6]. Finally, other methods work on the integrative analysis of networks built on each view by using an iterative optimisation analysis based on the local neighbourhood and then applying spectral clustering on the final integrated matrix [22].

We propose a network analysis based on the late integration approach that finds clusters by using two main steps: it first computes the hierarchical organisation of the samples in each view by means of a nested community detection technique and then it identifies the groups of samples that are consistent between the views. It is known that biological networks are highly clustered and have a small node-to-node distance [9]. Statistical network analysis tries to discover the underlying structure in these networks. Hierarchical structures, where vertices are divided into groups and sub-groups, are of particular interest since they build a hierarchy on the data and allow to easily understand the interactions between the nodes [10]. Efforts have been made to propose measures to quantify the level of hierarchy present in complex networks [13]. Unfortunately, many of these approaches work for directed [3] or undirected graphs where a rank of the nodes is known [20]. Our method can be used, not only to perform clustering but also to solve these problems. Indeed, it works on undirected graph structures where no prior information on the organisation of the samples structures is known and it is able to suggest the level of hierarchy present in the data and the structural organisation. We first performed a simulation study to prove the capability of the method to identify hierarchies in multi-view data, and then we applied it to the problem of multi-view patient sub-typing. Classical approaches to patient stratification are performed by studying differences in genetic association analysis by clustering patients through their gene expression data [24]. Indeed, high throughput technologies, such as microarrays and next-generation sequencing, have opened new possibilities for biomarker discovery and cancer sub-typing, by moving from single gene studies to an analysis encompassing the whole genome and/or transcriptome [7,11]. The application of multi-view clustering to multiple omic views, related to the same patients, can help identify groups of samples that share relevant molecular properties. Since diseases have complex phenotypes, the use of individual genes as biomarkers may not be effective in disease sub-typing. Moreover, multi-view techniques allow identifying the combinations of different molecular characteristics (i.e., genes, miRNAs, proteins) that best describe each sub-type [16, 19, 22].

We compared the patient multi-view clustering results of our approach with those recently obtained by two multi-view data integration methodologies, MVDA [16] and SNF [22]. MVDA works by factorising the cluster membership matrices obtained in every single view in a late integration approach. In this way, the information from different views is integrated at the result levels to obtain better performance with respect to the single view clustering algorithms. Similarity Network Fusion (SNF) is an intermediate multi-view clustering methodology that combines data of different types of genome-wide data (e.g. mRNA expression, miRNA, DNA methylation) to better characterise patient sub-groups. The integration is performed in an iterative manner, in such a way that, step after step, it updates the similarities between the patients with the information coming from the different views and leads to the construction of their interaction network. This network is used to cluster patients and to identify patient sub-types. We showed that the clustering impurity error obtained by our method is at least comparable with that obtained with the MVDA methodology.

The rest of the paper is organized as follows: in Sect. 2 our methodology is described. The Hierarchical block matrix is introduced along with our adaptation to use it for hierarchical clustering (Sect. 2.1). Furthermore, details on synthetic data generation and the simulation study are reported in Sects. 2.2 and 2.3, respectively. In Sect. 2.4 the multi-view data sets are described. In Sect. 3 results on the simulation study are reported and commented. In Sect. 4 the multi-view clustering results are discussed. Finally, concluding remarks are provided in Sect. 5.

2 Materials and Methods

The proposed methodology integrates the complementary information of different omic views to perform the multi-view clustering. It takes as input n matrices $M_i \in R^{F_i \times P}$ for $i \in 1, \ldots, n$, where F_i is the number of features and P is the number of samples. The result is a multi-view partitioning $C = \bigcup_{i=1}^{k}(C_i)$ of the samples information. The methodology is composed of four main steps as showed in Fig. 1: (i) similarity matrices construction: for each omic view the Pearson correlation between each couple of samples is computed; (ii) for each omic view the hierarchical block matrices are computed; the community detection algorithm used here is the Louvain modularity method, that is a greedy optimisation method that runs in $O(n \log n)$ [1] (iii) HBM matrix integration; (iv) Hierarchical clustering, with Ward minimum variance criterion [23], performed on the integrated matrix. The method was first tested on simulated data and then applied to real multi-view data for cancer patient sub-typing. The results are compared with recently proposed multi-view clustering methods such as MVDA [16] and SNF [22]. To make these results comparable with those already published in [16], the same number of clusters was used.

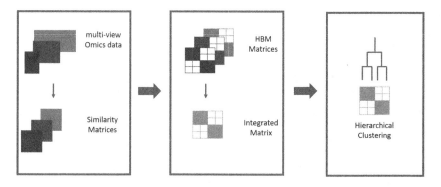

Fig. 1. Proposed methodology: (left) starting from the multi-view omic data (e.g. gene expression, miRNA expression, methylation) the samples similarity matrices are constructed by using the Pearson correlation metrics. (middle), for each similarity matrix, an HBM is constructed. The HBM matrices are then integrated into a single matrix that is used as a dissimilarity matrix for the hierarchical clustering method, used to identify the groups of samples (right).

2.1 Hierarchical Block Matrix (HBM)

In this work we applied the data integration algorithm proposed by Shavit et al. [18] (which we refer as HBM) to identify possible hierarchical structures in multi-view data sets. The method was designed to identify hierarchical and modular DNA structures within the cell nucleus, using different granularity of a chromatin fragment proximity measure. In this work, we investigated the potential of using the HBM method to identify hierarchical structures in multi-view data set and we applied it to the problem of identifying patient groups (modules) from multi-view genomic experiments for which the presence of hierarchies is not known a priori.

Let $N = (V, E)$ be an interaction network where $V = \{p_1, p_2, \ldots, p_n\}$ is the set of nodes representing the samples and $E = \{\{p_i, p_j\} | \rho_{p_i, p_j} \neq 0\}$ is the set of edges whose weights are computed as the Pearson correlation coefficients between the $i-th$ and $j-th$ samples. The HBM matrix is defined as follows: let C be the set of clusters in N, $C = \{c_l\}_{l=1}^{k}$ where $c_l \subseteq V$ and $k \geq 1$. We denote $B(1)$ to be an $m \times m$ matrix, with:

$$B(1)_{i,j} = \begin{cases} 1, & i, j \in c, c \in C, c \subseteq V, \\ 0, & \text{otherwise} \end{cases}$$

Let $N_1(V_1, E_1)$ be an undirected graph whose nodes are the clusters in N and $C_1 = \{c_{1,l}\}_{l=1}^{k_1}$ is the set of the clusters in N_1, with $c_{1,l} \subseteq V$ and $k_1 \geq 1$. Note that each cluster in C_1 is a union of sets (clusters in C) that contains nodes in V. Using a recursive definition, we denote N_s to be an undirected graph whose nodes are the clusters in N_{s-1} and $B(s)$ to be a $m \times m$ matrix, with:

$$B(s)_{i,j} = \begin{cases} 1, & i,j \in c, c \in C_{s-1}, c \subseteq V, \\ 0, & \text{otherwise} \end{cases}$$

where $C_{s-1} = \{c_{s-1,l}\}_{l=1}^{k_{s-1}}$ is the set of clusters in N_{s-1} with $c_{s-1,l} \subseteq V$ and $k_{s-1} \geq 1$, for $s \geq 2$. Note that if $B(s)_{i,j} = 1$ than for all $s' > s$, $B(s')_{i,j} = 1$ as well.

The hierarchical block matrix (HBM) of N is a non-negative symmetric $m \times m$ matrix, H, with: $H_{i,j} = min_s\{s | B(s)_{i,j} = 1\}$, for $s \geq 1$.

In this work, we adapted the hierarchical block matrix technique to identify the hierarchical structure in multi-view data sets and to better characterise patient sub-classes in omic data sets. In particular, for the patient sub-typing task, we applied the described method to construct the HBM matrices from the pairwise correlation matrices of the patients in each view, in order to identify the hierarchical structure of the patient classification. According to the definition given in [18], we merged the HBM matrices to obtain a stratification of the patients that is concordant in the different views.

Giving two HBMs A and B, the merging matrix M is defined as

$$M = \frac{H^A + H^B}{2}$$

where $M_{i,j}$ takes the "average level" between the levels at which i and j where assigned to the same community for the first time. The merged HBM $H^{A,B}$ can be obtained by merging their topologies.

Let $Z = \{z_i\}_{i=1}^{l_M}$ be the set of unique levels in M, sorted by increasing order, $H_{i,j}^{A,B}$ is given by:

$$H_{i,j}^{A,B} = g(M_{i,j})$$

where g is the function that takes a level in M and returns its index in Z such as $g(z_i) = i, z_i \in Z$.

Once the integrated HBM is computed, it can be used as input for a hierarchical clustering algorithm, to obtain the clusters of samples.

2.2 Synthetic Data Generation and Simulation Study

In order to test the ability of HBM to identify hierarchical structures in multi-view data sets, we performed different simulation studies. The simulation process is divided into three steps: (a) multi-view data set generation; (b) HBM algorithm execution; (c) evaluation of the HBM results; The multi-view data sets were generated in the following way: starting from two semi-positive covariance matrices S_1 and S_2 of size $N \times N$ (where N is the number of samples in the data set) we built a data set with two views by randomly sampling points from a multivariate normal distribution with mean equal to zero and covariance matrices S_1 and S_2. At the end of this process, we obtained two matrices $D_1 \in R^{M_1 \times N}$

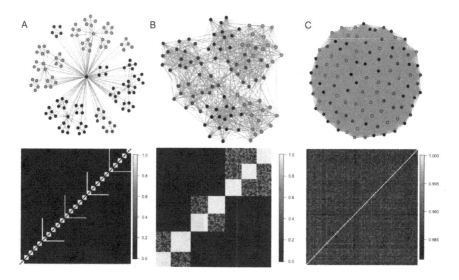

Fig. 2. Graphical visualisation of the synthetic datasets. The hierarchical networks, whose nodes are coloured based on the group they belong, underlying the synthetic datasets, are shown in the first row. The covariance matrices used to generate the synthetic datasets are shown in the second row. See text for differences between A, B, C cases. (Color figure online)

and $D_2 \in R^{M_2 \times N}$ where M_1 and M_2 were the number of features in the two different views, respectively, which, in these cases, were selected as 1000 and 500. The number of samples N was selected to be 100. Depending on the structure of the initial covariance matrices, different hierarchical data sets were generated. The first one was the nested hierarchical structure in which a pattern was nested and repeated into the network (See Fig. 2A).The second one was a nested block matrix structure that is a diagonal block matrix divided into blocks and sub-blocks (See Fig. 2B). The last one was a random network in which there was not a hierarchical structure (See Fig. 2C). For each experiment, a multi-view matrix was obtained by using the same covariance structure but different parameter settings, such as, for example, the number of blocks and sub-blocks. In the first experiment, the two nested hierarchical structures obtained nesting groups of 5 and 4 nodes, respectively. In the second experiment, in the first view, the number of blocks was three, each one composed by groups of two and three sub-blocks. While for the second view, the number of blocks was 2 each composed of two sub-blocks. In the third experiment, the two matrices were randomly sampled from a uniform distribution in the range $[0, 1]$. After the generation of the data set, the HBM algorithm was executed and a final merged multi-view hierarchical block matrix was obtained. Then, the purity index between the original clustering and the one obtained from the HBM3C matrix was computed in the following way:

$$purity(\Omega, C) = \frac{1}{N} \sum_k max_j |\omega_k \cup c_j|$$

where Ω is the original clustering, C is the multi-view clustering obtained from the HBM method, ω_k is the set of elements in the k-th cluster of Ω and c_j is the set of elements in the j-th cluster of C. Moreover, the Von Newman entropy [14] of the hierarchical block matrices and their histograms were investigated. Indeed, the Von Neumann entropy, which is the natural extension of the Shannon information entropy, is a widely adopted descriptor to measure the mixedness of a system and we used it as an indicator of the hierarchical structure present in the HBM matrix.

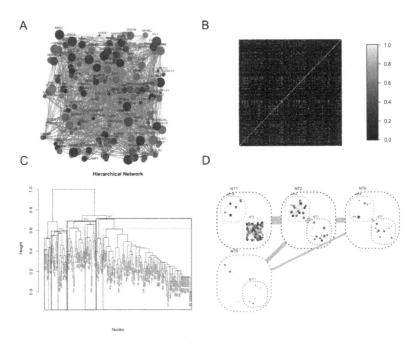

Fig. 3. The network of co-expression genes used in the study; see text for a description of the network (A). Adjacency matrix (B). Hierarchical clustering applied to the gene co-expression matrix to identify groups of modules (C). Hierarchical visualisation of the network performed with the RedeR package (D).

2.3 Simulation Based on a Gene Co-expression Matrix

To show the capability of the HBM methods to identify hierarchical and modular structures in a network, a further experiment based on a real gene co-expression matrix was performed. Since, gene co-expression networks are known to have hierarchical and modular structure [15], they are well suited to our purposes. The gene expression data used in this study comes from a study on the regulation effect of the estrogen receptor in breast cancer [4]. The gene co-expression network is available in R through the RedeR bioconductor package [5]. The network, showed in Fig. 3, contains 174 early ER-responsive genes connected by

1061 edges representing the reliable correlation based connections between the genes. As per the simulated data set, the co-expression network was treated as a covariance matrix to generate two multi-view data sets of 174 objects, randomly sampled by a multivariate normal distribution with zero mean. The number of features is selected as 1000 and 500 in the two different views. A hierarchical clustering was applied to the network in order to identify the modules that will be later used to compute the purity index between the clusters identified by the HBM algorithm and the real hierarchical organisation in the network. Also in this setup, the experiment were performed adding different amounts of noise to the original network in order to study the capability of the HBM method to identify hierarchies in presence of noise.

2.4   Multi-view Data Set for Patient Sub-typing

The HBM method was applied to two multi-view omic data sets (see Table 1) related to Glioblastoma and Prostate Cancer, respectively.

The data set TCGA.GBM was downloaded from The Cancer Genome Atlas (TCGA) (https://cancergenome.nih.gov/). The data set is composed of 167 patients represented by two views: gene expression and miRNA expression. As described in [21], the patients were divided into four classes: Classical, Mesenchymal, Neural and Proneural. The MSKCC.PRCA data set was downloaded from the Memorial Sloan Kettering Cancer Center data portal (http://cbio.mskcc.org/cancergenomics/). It has four views: clinical data, gene expression, miRNA expression and copy number variation. Patients were classified into two classes by using the tumour stage. All the downloaded data sets were already preprocessed. As a further step, features with low variance were eliminated and batch effect removal was performed.

Table 1. Multi-view omic data sets used in the experiments. The TCGA.GBM data set has two views that are the gene and miRNA expressions, while the MSKCC.PRCA data set has three views that are the gene and miRNA expressions and the Copy Number Variation (CNV).

Data set	Nr. Views	N. Samples	N. Classes	Disease
TCGA.GBM	2	167	4	Glioblastoma
MSKCC.PRCA	4	88	2	Prostate cancer

3   Simulation Study Results

In order to prove the ability of the hierarchical block matrix algorithm to retrieve hierarchical clustering structures from multi-view data sets, four different simulation studies were performed, as explained in the previous section. In the first one a nested hierarchical multi-view data set was generated, while, in the second one, a nested block matrix was created. In the third one, a random multi-view data set was generated (see Fig. 2). Furthermore, a multi-view data set was generated based on a real gene co-expression matrix (see Fig. 3). Then, the HBM algorithm was executed and the multi-view hierarchical clusterings were retrieved, while the purity index between the known clustering and the new one was computed. For each simulation setting, the experiment was repeated 10 times by changing the percentage of noise injected in the original hierarchical covariance matrices. As we can see from Table 2 the HBM algorithm is able to retrieve hierarchical structures when present in the data, as for the block matrix, the nested hierarchical organisation and the data set based on the gene co-expression matrix. In these cases, the purity index is always higher than 64%. Of course, the lower the noise the higher is the purity index. On the other hand, in case of no hierarchical structure in the data set, the purity index is much lower. The hierarchical block matrices obtained with no noise are reported in Fig. 4. The HBM matrices contain integer numbers that indicate in which level of the hierarchical agglomeration algorithm two samples were put in the same cluster. Visual inspection of the HBM matrices and their histograms can suggest if a hierarchy is present or absent in the data. Indeed, in the case of a random sample data set, the HBM matrix is full of 1, meaning that all the samples are clustered at the first iteration of the algorithm (Fig. 4C). On the other hand, different values are retrieved in the hierarchical organisation. Furthermore, the Von Neumann entropy of the multi-view HBM matrices was measured. The values are of 258.17 for the nested hierarchical experiment, 352.75 for the nested block matrix structure, 503.96 for the gene co-expression based structure and 1101.68 for the random network experiment. These values correspond to the experiments with no noise in the synthetic generated data sets. These results suggest Von Neumann entropy could be used as an indicator of how much hierarchical structure is present in the HBM matrix since it has lower values in case of well defined structures and higher values in case of random structures.

4   Application to a Real Data Set

The proposed methodology was applied to two multi-view omic cancer data sets to identify patient sub-typing. For each of the omic views available in the data sets an HBM matrix was computed. The single HBM matrices were then integrated with the hierarchical agglomerative strategy previously described. Finally, on the integrated matrix a hierarchical clustering was performed. The obtained clustering was compared with those coming from two state-of-the-art methodologies called MVDA [17] and SNF [22]. The comparison was performed in terms of purity of the clusters with respect to the known patient subclasses. The number of clusters used in the analysis is that previously used in the MVDA paper [17]. As reported in Table 3, the accuracy obtained by the integrative methods (MVDA, SNF, and HBM) is higher or comparable to the accuracy obtained by the single views. The accuracy obtained by HBM is comparable to those obtained with the MVDA and SNF approaches in the case of the MSKCC data set, while it is significantly higher than MVDA and SNF in the case of the TCGA.GBM data set.

Table 2. Purity index values between the known clustering structure and the multi-view ones obtained with the HBM methods by varying the amount of noise in the data (the x value in $Th_x\%$ means how much is the percentage of noise in the generated covariance matrices. In the row names, BM stands for the block matrix experiment, NS stands for recursively nested experiment, RD stands for the random matrix experiment and CE stands for the gene co-expression experiments. The $_1$ or $_2$ close to the names on the rows indicate if the comparison was performed with respect to the labelling of the samples in the first or second view. Purity values close to 1 indicate better results.

	Th_0%	Th_1%	Th_2%	Th_3%	Th_4%	Th_5%	Th_6%	Th_7%	Th_8%	Th_9%
BM_1	0.95	**0.99**	0.82	0.81	0.71	0.71	0.81	0.70	0.79	0.69
BM_2	**1.00**	0.99	0.98	0.97	0.97	0.95	0.90	0.88	0.85	0.80
NS_1	0.88	0.88	**0.89**	0.87	0.84	0.85	0.86	0.81	0.70	0.64
NS_2	**0.97**	0.88	0.88	0.86	0.88	0.89	0.89	0.88	0.65	0.64
RD_1	**0.49**	0.46	0.40	0.38	0.37	0.39	0.38	0.36	0.35	0.33
RD_2	**0.55**	0.53	0.50	0.48	0.46	0.43	0.40	0.38	0.37	0.32
CE_1	**0.75**	0.72	0.71	0.70	0.66	0.62	0.59	0.59	0.57	0.55
CE_2	**0.73**	0.71	0.71	0.69	0.64	0.63	0.57	0.55	0.55	0.53

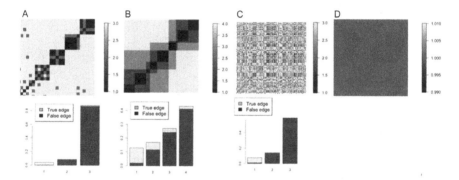

Fig. 4. HBM matrices along with their histograms. The HBM associated to the nested hierarchical model has three levels of hierarchy as shown by its associated histogram (A), while the HBM of the nested block matrix has four levels of hierarchy as shown by its associated histogram (B). Also, the HBM based on the gene co-expression matrix shows three levels of hierarchy (C). On the other hand, the HBM associated to the random data set does not show hierarchical structures, indeed the whole matrix is filled by ones (D). The histograms are filled by the number of true edges (light grey) and false edges (dark grey) present in the original matrix. The true edges are mainly enriched in the first level of the hierarchy, meaning that the samples connected by true edges are clustered together in the first level of the HBM methodology. On the other hand, most of the second, third and fourth levels are full of false edges because these samples are further away from each other and are merged together later by the HBM method. (Color figure online)

In term of computational time, the single view hierarchical clustering method is the fastest one. Between the integrative methods, computational time of the HBM and SNF are comparable. The computational time of the MVDA method is significantly higher than the HBM method.

Table 3. Clustering purity measures are reported for both single view clustering (applied to gene expression, miRNA expression, and CNV), and multi-view clustering approaches (MVDA, SNF, and GBM). Please note that TCGA.GBM matrix has only two views (gene expression and miRNA expression), while the MSKCC has three views (gene and miRNA expression and CNV). The clustering purity measure is significantly higher in the multi-view clustering obtained by the HBM method in both the glioblastoma data set (TCGA.GMB) and prostate cancer data set (MSKCC.PRCA). The single view clusterings are obtained by using the hierarchical clustering algorithm on each view.

Approach	TCGA.GBM	MSKCC
Gene expression	79.05%	63.64%
miRNA expression	61.08%	60.23%
CNV	-	36.36%
MVDA	63.48%	63.64%
SNF	76.00%	64.00%
HBM	**83.24%**	**64.78%**

5   Conclusion

In this study, we proposed an adaptation of the Hierarchical Block Matrix method [18] to the problem of multi-view clustering with a particular application to patient sub-typing.

We first demonstrated the effectiveness of the HBM method to identify hierarchical structures in networks, derived from multi-view data sets, by performing four different simulation studies. We computed the normalized mutual information between the known clustering structure imposed in the simulated multi-view data set with that obtained with the HBM method. When the hierarchy is present in the data set, the mutual information is never lower than 0.60 even with an injected noise of 90%.

Then we applied our method to two real multi-view omic data sets related to cancer studies to identify patient sub-typing. Based on the pairwise Pearson correlation between the patients, an HBM matrix is built for each view of the data sets, and then the HBM matrices are merged into a single one that is used as starting point for a hierarchical clustering analysis.

We compared the clustering purity scores obtained with the HBM based method with those obtained with the MVDA and SNF methodologies. Our results show that all multi-view approaches perform slightly better than the single view hierarchical clustering. Moreover, the HBM clustering reaches higher performances compared to SNF and MVDA. Altogether, the HBM method is a viable alternative methodology for multi-view patient sub-typing.

References

1. Blondel, V.D., Guillaume, J.L., Lambiotte, R., Lefebvre, E.: Fast unfolding of communities in large networks. J. Stat. Mech. Theory Exp. **2008**(10), P10008 (2008)
2. Cai, X., Nie, F., Huang, H.: Multi-view k-means clustering on big data. In: IJCAI, pp. 2598–2604 (2013)
3. Carmel, L., Harel, D., Koren, Y.: Drawing directed graphs using one-dimensional optimization. In: Goodrich, M.T., Kobourov, S.G. (eds.) GD 2002. LNCS, vol. 2528, pp. 193–206. Springer, Heidelberg (2002). https://doi.org/10.1007/3-540-36151-0_19
4. Carroll, J.S., et al.: Genome-wide analysis of estrogen receptor binding sites. Nat. Genet. **38**(11), 1289 (2006)
5. Castro, M.A., Wang, X., Fletcher, M.N., Meyer, K.B., Markowetz, F.: Reder: R/bioconductor package for representing modular structures, nested networks and multiple levels of hierarchical associations. Genome Biol. **13**(4), R29 (2012)
6. Chaudhuri, K., Kakade, S.M., Livescu, K., Sridharan, K.: Multi-view clustering via canonical correlation analysis. In: Proceedings of the 26th Annual International Conference on Machine Learning, pp. 129–136. ACM (2009)
7. Ciriello, G., Miller, M.L., Aksoy, B.A., Senbabaoglu, Y., Schultz, N., Sander, C.: Emerging landscape of oncogenic signatures across human cancers. Nat. Genet. **45**(10), 1127–1133 (2013)
8. Greene, D., Cunningham, P.: A matrix factorization approach for integrating multiple data views. In: Buntine, W., Grobelnik, M., Mladenić, D., Shawe-Taylor, J. (eds.) ECML PKDD 2009. LNCS (LNAI), vol. 5781, pp. 423–438. Springer, Heidelberg (2009). https://doi.org/10.1007/978-3-642-04180-8_45
9. Grigorov, M.G.: Global properties of biological networks. Drug Discov. Today **10**(5), 365–372 (2005)
10. Herlau, T., Mørup, M., Schmidt, M.N., Hansen, L.K.: Detecting hierarchical structure in networks. In: 2012 3rd International Workshop on Cognitive Information Processing (CIP), pp. 1–6. IEEE (2012)
11. Kandoth, C., et al.: Mutational landscape and significance across 12 major cancer types. Nature **502**(7471), 333–339 (2013)
12. Liu, J., Wang, C., Gao, J., Han, J.: Multi-view clustering via joint nonnegative matrix factorization. In: Proceedings of the 2013 SIAM International Conference on Data Mining, pp. 252–260. SIAM (2013)
13. Mones, E., Vicsek, L., Vicsek, T.: Hierarchy measure for complex networks. PloS one **7**(3), e33799 (2012)
14. Petz, D.: Entropy, von Neumann and the von Neumann entropy. In: Rédei, M., Stöltzner, M. (eds.) John von Neumann and the Foundations of Quantum Physics. VCIY, vol. 8, pp. 83–96. Springer, Dordrecht (2001). https://doi.org/10.1007/978-94-017-2012-0_7
15. Ruan, J., Dean, A.K., Zhang, W.: A general co-expression network-based approach to gene expression analysis: comparison and applications. BMC Syst. Biol. **4**(1), 8 (2010)

16. Serra, A., Fratello, M., Fortino, V., Raiconi, G., Tagliaferri, R., Greco, D.: MVDA: a multi-view genomic data integration methodology. BMC Bioinform. **16**(1), 1 (2015)
17. Serra, A., Fratello, M., Greco, D., Tagliaferri, R.: Data integration in genomics and systems biology. In: 2016 IEEE Congress on Evolutionary Computation (CEC), pp. 1272–1279. IEEE (2016)
18. Shavit, Y., Walker, B.J., et al.: Hierarchical block matrices as efficient representations of chromosome topologies and their application for 3C data integration. Bioinformatics **32**(8), 1121–1129 (2016)
19. Taskesen, E., et al.: Pan-cancer subtyping in a 2D-map shows substructures that are driven by specific combinations of molecular characteristics. Sci. Rep. **6**, 24949 (2016)
20. Trusina, A., Maslov, S., Minnhagen, P., Sneppen, K.: Hierarchy measures in complex networks. Phys. Rev. Lett. **92**(17), 178702 (2004)
21. Verhaak, R.G., et al.: Integrated genomic analysis identifies clinically relevant subtypes of glioblastoma characterized by abnormalities in PDGFRA, IDH1, EGFR, and NF1. Cancer Cell **17**(1), 98–110 (2010)
22. Wang, B., et al.: Similarity network fusion for aggregating data types on a genomic scale. Nat. Methods **11**(3), 333–337 (2014)
23. Ward Jr., J.H.: Hierarchical grouping to optimize an objective function. J. Am. Stat. Assoc. **58**(301), 236–244 (1963)
24. West, M., et al.: Predicting the clinical status of human breast cancer by using gene expression profiles. Proc. Natl. Acad. Sci. **98**(20), 11462–11467 (2001)

Author Index

Printed in the United States
By Bookmasters